WALENTY ROŹDZIEŃSKI

OFFICINA FERRARIA

This is Number 9 in a series of monographs in the history of technology and culture published jointly by the Society for the History of Technology and The MIT Press. The members of the editorial board for the Society for the History of Technology Monograph Series are Melvin Kranzberg, Cyril S. Smith, and Thomas P. Hughes.

Previous publications in the series include:

History of the Lathe to 1850, Robert S. Woodbury

English Land Measuring to 1800: Instruments and Practices, A. W. Richeson

The Development of Technical Education in France 1500-1850, Frederick B. Artz

Sources for the History of the Science of Steel 1532-1786, Cyril Stanley Smith

Bibliography of the History of Technology, Eugene S. Ferguson

The Hollingworth Letters: Technical Change in the Textile Industry, 1826-1837, Thomas W. Leavitt, editor

Science in France in the Revolutionary Era, Described by Thamas Bugge, Danish Astronomer Royal and Member of the International Commission on the Metric System (1798-1799), Maurice P. Crosland, editor

Moving the Obelisks, Bern Dibner

WALENTY ROŹDZIEŃSKI

OFFICINA FERRARIA

A Polish poem of 1612
describing the noble craft
of ironwork

Translated by Stefan Płuszczewski

Edited by Wacław Różański
and Cyril Stanley Smith

Introduction by Jerzy Piaskowski

Published jointly by the
Society for the History of Technology
and The MIT Press
Cambridge, Massachusetts, and London, England

This book was printed and bound by The Colonial Press Inc. in the United States of America.

Library of Congress Cataloging in Publication Data

Roździeński, Walenty, ca. 1560-ca. 1622.
Officina ferraria.

(Monograph series—Society for the History of Technology; no. 9)
Bibliography: p.
Includes index.
1. Iron—Metallurgy—Poetry. I. Title.
II. Series: Society for the History of Technology Monograph series; no. 9.
PG7157.R6503513 1977 891.8'5'13 76-26592
ISBN 0-262-18079-0

CONTENTS

PREFATORY NOTE

Cyril Stanley Smith

The seventeenth-century Polish work here published in English for the first time is in a form unusual for technical literature. Although written in the vernacular by a working smith specifically to record and display the pride and problems of his craft, it is a somewhat elegant poem, embellished with an extensive "historical" introduction based on ancient and mythical sources in self-conscious emulation of traditional literary works. It can be classed neither with the earlier popular German *Kunstbüchlein* (pamphlets on dyeing, on assaying, on handling steel, and on many other craft and domestic "secrets") nor with the carefully written technological treatises of Biringuccio and Agricola, nor with the compilations made by contemporary literary scholars—but it does have something in common with all of these. It is particularly important because it reflects the values of rural workers and is not the work of an academic or of someone in a center of commerce, in the church, or in government.

The poem is of particular interest as far as metallurgical history is concerned because it treats of iron, which, despite the overwhelming dominance of its properties and utility, has left scant written record, supposedly because the low social position of its skilled but sooty workers discouraged interest at the educated professional level. This is probably related to the more concentrated capital investment in copper mining and smelting and to the nature of the raw materials and processes (including desilverization), which lent themselves to the quasi-scientific study and control that was justified by the value of the products.

The seventeenth century saw intense activity in physics and chemistry but, with the exception of Mathurin Jousse's *Fidelle Ouverture de l'Art de Serrurier* (1627) and A. A. Barba's account of silver extraction in the New World in his *Arte de los Metales* (1640), there was little writing on metals beyond reprints and plagiarism of the great sixteenth-century writings.

This changed in the eighteenth century, but even R. A. F. de Réaumur in his great treatise of 1722, which is the first book-length treatment of the metallurgy of iron, has little to say on common smelting practice. Good documentation on iron smelting begins a century and a half after the *Officina Ferraria* with Emanuel Swedenborg's *De Ferro* in 1734, followed by the *Art des Forges et Fourneaux à Fer* edited by Gaspard Courtivron and E. J. Bouchu, published in 1762 by the Académie des Sciences as part of the monumental *Descriptions des Arts et Métiers.* Thereafter Grignon, Jars, Rinman, Hassenfratz, Karsten, and subsequent authors provide a continuing reasonably detailed picture of the technology of iron.

Knowledge of the smelting of iron prior to the eighteenth century therefore comes mainly from brief notes in nontechnical works to which the topic is incidental. The significant sources are all listed in Prof. Piaskowski's bibliography (pp. 103-104). He points out that works like Bourbon's "Ferraria" and the poems of Villichius and Schroeter on the salt mines at Wieliczka (1543 and 1563, respectively) are intended to give a human picture of work and therefore constitute a special kind of Renaissance poetry. The *Officina Ferraria* is in a different class. Roździeński was a smith himself, and he wrote of his daily work and experience. He was the master of a forge and knew at first hand the problems of running an enterprise and managing skilled and unskilled personnel. Except for the fanciful introduction, the poem reflects a sense of reality and immediate experience that all secondary literature necessarily lacks, and because of this it is of particular value to the historian.

Though a practical man, Roździeński obviously was educated and was trying to conform to the pattern set by learned writers of his day. The extensive citations of the classical literary sources, traditional and uncritical though they are, seem meant to convey an air of erudition, while the versified form, an unusual clothing for the subject, suggests that the work was intended for a popular audience. Despite the selection, suppression, and emphasis that was undoubtedly necessary to fit the poetic meter (much like the distortion of technical details of mines, forges, and smelting houses when used as themes in paintings), the end result was the preservation of a useful record from a period when more purposeful technical prose was entirely lacking.

The *Officina Ferraria* is neither unique nor particularly impressive as art or formal literature, but it is nevertheless of great value as a record of human experience of a kind not commonly studied. Only in the last few decades have Western scholars come to appreciate the level of history reflected here, the personal and microsocial experience of the technologist as he makes the materials and goods upon which economic and political institutions are built. Roździeński's poem discusses the qualities of minerals from different localities and reflects the discovery of those properties and reactions of matter upon which later scientific theories would be based. It conveys rather vividly a sense of the interrelatedness of forests, mountains, and rivers, with the work of the charcoal burners and miners culminating in the smelting house set up beside a river for power. It discusses the rise and fall of individual forges in recent history, providing evidence for the migration of peoples, with comments on the antiquity of the workings based on the size of slag piles. Ironworking was no longer the sole province of individual craftsmen, but involved investment and organization on a scale that foreshadows later capitalistic enterprise.

The poem details the assignment of special responsibilities and duties to individual men, together with the inducements and restraints necessary to maintain a productive organization. Some men, like the charcoal handler, have supervisory duties of a scale and diversity considerably beyond that implied by their title; others are simply workers with few needs beyond housing, food, and especially drink. The poem records pride, skill, and mean thievery in addition to both occupational health hazards and the joy of productive work. It retells the lore of past discoveries of fabulous mines, of the exhalations that appear above ore bodies and lead men to new findings, and of the gnomes and elves that help and hinder work in the mine and forge.

The layout of the whole establishment is well described: the bloomery hearth with its waterwheel, bellows, charcoal lining, and carefully shaped tuyeres, the heavy water-driven hammer of the forge, and the many smaller tools necessary for the work. Throughout there are admonitions on the handling of different kinds of materials in accordance with their particular qualities, of which only a practical man could be aware. The annual cycle of cutting wood and burning charcoal in the forest is particularly well described because of its immediate importance to the operation of the forge, as is the nature of the different kinds of ore and their behavior in smelting, but there is very little on the actual processes of mining, with which the forgemaster would not necessarily have been intimately involved. Similarly, Roździeński's interest in the product ends with semifabricated forms of the primary forge, and there is very little indeed on the final shaping and use of iron by craftsmen or other industries.

Reverting to the traditional tone of the introduction, the final section is a song of praise to the virtues and utility of iron. This erroneously assumes that all early tools were made of iron, and was obviously written by a producer of the metal, not a user of it. Indeed, this end-product aspect of metallurgy attracted remarkably little attention from writers of any kind prior to the time when it became the physical metallurgy of the early twentieth century and the materials engineering of today. The very real knowledge that did exist was mainly in the minds of illiterate craftsmen who worked individually in an unorganized fashion without needing large amounts of capital or producing problems or profits enough to attract the interest of either the learned or the aristocracy.

Only one copy has survived of the original printing of the *Officina Ferraria* in 1612, and even this was unknown until 1929. Its rediscovery created much interest in Poland—more, perhaps, as an indication of early national culture and for its value to students of the history of the Polish language than for its metallurgical content—and it has not yet become incorporated in metallurgical or literary scholarship outside the country of its origin. In the course of translation into another language it is inevitable that many of the values that are of particular interest to Polish scholars will be lost. The technical and economic facts and nuances that remain are; however, more than enough to justify the effort of translation and study by historians of any nationality.

The writer first encountered the *Officina Ferraria* in 1965 when in Poland on the occasion of the Eleventh

International Congress on the History of Science. References to *grąpie* made in a talk by Prof. Radwan suggested a connection (now shown by Prof. Piaskowski to be erroneous) with the little lumps of cast iron found in ancient smelting sites in the Holy Cross Mountains, which in turn seemed to be of importance in connection with the carburization of steel *à la Chinois* and the production of granules in the first description of a European blast furnace (see *Technology and Culture,* vol. 5 [1964], pp. 386-407). It was tantalizing to learn of a rich source of metallurgical history in a language that I could not read! At last, thanks to the labors of Stefan Płuszczewski, it is available to all readers of English, placed in proper perspective by the foreword of Prof. Radwan and the introduction of Prof. Piaskowski.

It is hoped that the translation will serve not only to convey an understanding of the production of iron and the nature of industrial organization before the introduction of the blast furnace changed both forever, but also as a small contribution to international understanding and scholarly friendship.

FOREWORD

Mieczysław Radwan

In 1929 a brief bibliographic notice by Leon Formanowicz announced that a copy of a previously unknown 92-page poem celebrating "the noble craft of ironwork" had been found in the library of the Gniezno chapter house, bound together with other pamphlets of the sixteenth and seventeenth centuries. Circumstances revealed later indicated that the copy of the book had come from the library of Maciej Grabski, who lived in the latter part of the seventeenth century.

This announcement was a great surprise, as neither the poem nor the author had previously been mentioned by any Polish bibliographer. No other copy of the work has been found up to the present time.

The first, abridged, edition of 1933, prepared by Roman Pollak, Professor of the University of Poznań, caused an understandable sensation. In 1936 Prof. Pollak published an unabridged edition with brief explanations; next came a reprint in 1948 and finally the fully documented edition in 1962.

The first publication of the poem initiated a lively discussion among historians of science and of technology, both in Poland and abroad. Since then the poem and its author have been referred to in nearly every Polish scientific periodical; no historiographical paper concerning the seventeenth century could fail to note the work of the Silesian poet and ironworker or to recognize its significance in the history of Polish culture. Today the bibliography of studies involving Roździeński's work comprises more than five hundred items which have appeared in learned and popular publications, textbooks, encyclopedias, etc.

The full bibliography of Roździeńściana by W. Piasecki et al. (entry d37 in the bibliography of this book) shows 123 papers published in the decade following the discovery of the poem, 60 in the next decade (including the years of the Second World War), and 150 from 1950 to 1959. About a hundred papers appeared in the two years 1962 and 1963, the last years included in the bibliography.

The work has taken a very important place in historical studies of the literature and sociology of Poland, quite apart from its interest in connection with metallurgical history, which is the main concern of the publishers of the present translation. It has been studied not only as literature but also as a source of documentary evidence concerning the period of its writing.

Sociologists were interested in the fact that ironworkers existed as a separate social group within the community. Geographers found ample material relating to Silesian topography. Ethnologists found silhouettes of Renaissance miners and smelters. Linguists—especially polonists—found rich material not only in the field of the ironworkers' professional language, but also in the field of literary Polish of the sixteenth and seventeenth centuries. And historians of technology found an excellent and trustworthy picture of the techniques for the smelting and forging of iron used in those days.

Finally, Polish scholars acquired important additional evidence of the cultural unity of the Silesian region and of the territories of the Polish Kingdom of those days—despite the fact that, owing to the unhappy course of history, this region had lain for more than three centuries within the boundaries of a foreign state which discouraged any manifestation of Polish culture.

TRANSLATOR'S NOTE

Stefan Płuszczewski

The language in which Walenty Roździeński (pronounced Rose-*then*-ski) wrote three and a half centuries ago contained many words and idioms that are absent from present-day Polish, so that the original version of the poem is by no means easy to read. Some obsolescent words remain in the English and serve to give an appropriately archaic flavor to the poem, but in general the English-speaking reader will now have easier access to Roździénski's thought than does the present-day Pole.

The task of the translator consisted in transposing into English a Polish text of the early seventeenth century, characterized by:

1. numerous deviations from the commonly used Polish language of the epoch, including some traces of the effects of the Czech language, which was for a long time the official language in Silesia;

2. many technological names and descriptions whose content it has proved difficult (and sometimes impossible) to unravel;

3. the author's inclination to use certain words and phrases too frequently (according to the index of words in the 1962 Polish edition, the author used the words "to be" 357 times, "which" 329 times, "he who" 426 times, "this" 432 times).

The translation required both an understanding of the original Polish text and a knowledge of the comparable English technological terminology. The works of W. Grier,[1] J. M. Swank,[2] and H. R. Schubert[3] have been of assistance to the translator in the latter respect.

A few specific points about the translation should be mentioned. The works in which iron was smelted was equipped with a bloomery furnace or hearth and a hammer for the initial forging of a bloom. Roździeński called this works *kuźnica,* translated as "forge," and the person (the tenant of the owner) who kept it *kuźnik,* "forgemaster." The latter word is also sometimes applied to simple blacksmiths and coppersmiths, and in a few cases, where the context clearly demands it, the simpler term "smith" has been used in the translation.

1. Entry c16 in the bibliography of this book.

2. c51.

3. c48.

In some verses the author refers to the forge as *oficyna,* from the Latin, and this has been translated as "workshop." To designate the whole building containing the forge Roździeński uses the word *huta,* translated as "smelting house."

The foreman who ran the smelting in the bloomery was called a bloomer (in Polish, *starszy dymarz* or *starszy smelcyrz*).

In discussing the water drive (folio 12v, line 20) Roździeński uses the word *zawod.* This has been translated "driving force," though "driving moment" would have been more appropriate in modern terminology. The latter term is, however, not to be found in the chapter on waterwheels in Grier's handbook, published in 1835, and was therefore thought inappropriate here. In another Slavic language, Russian, one of the meanings of *zawod* is "winding," as of a clock.

In translating the names of the implements used in the forge I found the compilation of W. Bastyr and E. Paszkowski[4] to be quite useful. These tools are further discussed in Prof. Piaskowski's introduction, and a glossary of translations used for technological terms may be found in the rear of the book.

The ease with which Roździeński discusses technological matters is notable. For example, he recommends erecting a forge on a watercourse with a high head "so that the waterwheels do not wade." This expression is both clever and graphic.

The poem has been translated into prose. The division into stanzas has, however, been maintained. This has produced some inevitable variations in length and clumsy transitions, but will allow the interested reader to refer back easily to the original. In addition, the beginning of each leaf and its verso of the original edition is noted.

Note on the Format

Passages that appeared in the original in Latin have been set off in lighter type in this translation. Marginalia—usually consisting of simple citations, in

4. c4.

Latin, of the source from which a piece of information has been culled, though sometimes more informative—have been retained where they do not repeat information given in the text. At a few points where it seemed appropriate, they have been interpolated into the text. Notes introduced by the translator or the editors are marked off by brackets and set in italic type.

ACKNOWLEDGMENTS

This American edition of Walenty Roździeński's Polish metallurgical poem *Officina Ferraria* has been prepared as the result of an agreement between the Institute of the History of Science and Technology of the Polish Academy of Sciences and the Society for the History of Technology, through the initiative of the representatives of these institutions, Prof. Jerzy Piaskowski, the late Prof. Mieczysław Radwan, Prof. Eugeniusz Olszewski, Dr. Melvin Kranzberg, and Prof. Cyril Stanley Smith, who first proposed the project.

Our acknowledgments are due first of all to Prof. Bogdan Suchodolski, Head of the Institute of the History of Science and Technology of the Polish Academy of Sciences, which financed the preparation of the manuscript, and to Prof. Eugeniusz Olszeswki, Head of the Section of Technological Science of the Institute. The project was carried out by the Section on the History of Polish Metallurgical and Foundry Technology of the Polish Academy of Sciences, acting in Krakow on behalf of the above Institute. Prof. Wacław Różański, Chief of the Section, supervised the project and Mr. Stanisław Miczulski coordinated the editorial work. The text was translated by Stefan Płuszczewski, who unfortunately died before the work was finished. The translation was kindly verified by Mrs. Claire Grece-Dąbrowska and Mrs. Celia Beven-Oyrzanowska, and a few changes recommended by the advisors (especially Prof. Piaskowski) have been introduced by Miss Maria Klimosch. Many small changes in English wording were made by the American editor. Our thanks are also due to Dr. Stefan Zmuda for preparing the maps.

Finally, acknowledgment is due to the late Prof. Roman Pollak, who in 1929 saw the technological significance of the *Officina Ferraria* and later prepared the modern critical editions in Polish on which this translation is based.

The Society for the History of Technology gratefully acknowledges a grant from the United States Steel Foundation toward the cost of publication.

INTRODUCTION

Jerzy Piaskowski

I. Walenty Roździeński and the Situation of Forgemasters in the Kingdom of Poland and Silesia in the Late Sixteenth Century

Walenty Roździeński was a forgemaster. He never held any position of national importance, and for this reason there is but little information concerning him on the public record. Archival research has yielded some data, but it is not at all certain whether all the references found are to the author of the *Officina Ferraria.*

Certain details of Roździeński's life may be found in his poem. There he states that an ancestor of his named Hercyg–possibly mythical–was a forgemaster from Meissen who settled in Silesia, bringing with him the knowledge of improved methods of iron smelting (folio F2). Archival sources indicate that this move probably took place early in the sixteenth century.[1]

From this progenitor's name (*herzig* [Ger.] = *serdeczny* [Pol.] = hearty) Roździeński traces the heraldic device "Serdecznik," though it may well have happened the other way round, the name Hercyg being derived from the armorial design. At any rate, no such arms are to be found in Polish heraldic sources, so it is quite likely that they were invented by the author.[2]

Master Walenty's ancestors changed their name to Brusek, presumably after the locality in Silesia where they ran a forge, it being the custom that a son, on inheriting his father's forge, also assume the name of the locality. Hence Master Walenty, after taking over the forge of his father Jakub Brusek at Roździeń, adopted the name Roździeński. (See Figures 1 and 2 for the location of forges and towns mentioned in the poem.)

The poem traces the name "Brusek" back to the Roman name Brutus. This was no doubt a fantasy, typical of the period: a forgemaster from the environs of Lubliniec could not have had anything in common with the Romans. In Polish the word *brusek* means a grinding wheel or whetstone, used by smiths.

1. See page 294 of entry 29 in part d of the bibliography of this book (hereinafter d29:294).

2. a1936:XVIII.

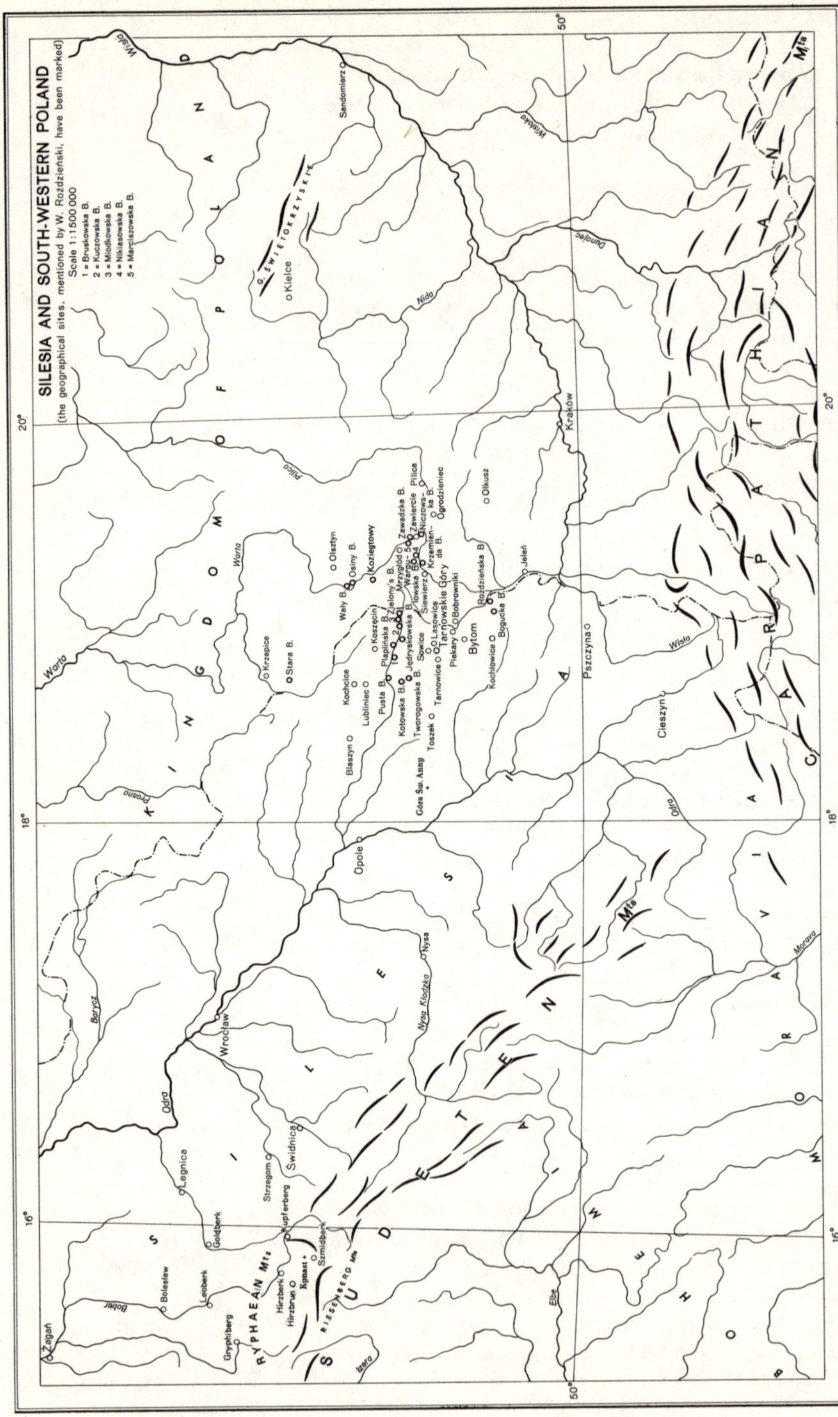
SILESIA AND SOUTH-WESTERN POLAND
(the geographical sites, mentioned by W. Rożdzieński, have been marked)
Scale 1:1 500 000
1 = Bruskowska B.
2 = Kuczowska B.
3 = Mlodkowska B.
4 = Nikiasowska B.
5 = Marciszowska B.
KINGDOM OF POLAND
SILESIA
SUDETEN Mts
CARPATHIAN Mts
MORAVIA
BOHEMIA
Kraków
Wrocław
Opole
Bytom
Tarnowskie Góry
Kielce
Sandomierz
Pszczyna
Cieszyn
Legnica
Świdnica
Nysa
Odra
Wisła
Warta
Pilica
50°
20°
18°
16°

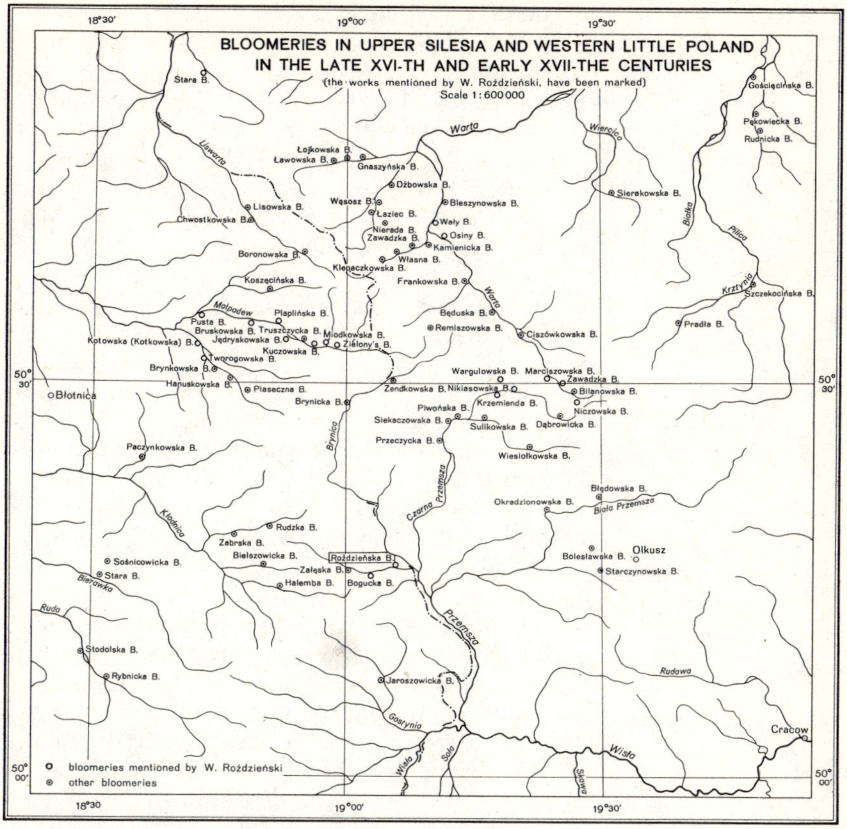

Figure 2 Bloomeries in Upper Silesia and western Little Poland in the late 16th and early 17th centuries. The works mentioned in the poem are marked.

Figure 1 Map of Silesia and southwestern Poland. The geographical sites mentioned in the poem are marked.

All these fictions indicate a good deal of concealed irony toward the upper classes of society. Both the tracing of the name Brusek to Brutus and the fictitious coat of arms "Serdecznik" may suggest that Hercyg, ancestor of the Roździeński family, was also invented by Master Walenty. Perhaps by creating progenitors among forgemasters from Meissen the author of the *Officina Ferraria* thought to cover himself with a certain splendor.

In any case, the city archives at Lubliniec contain mention that in 1562 Walenty Brusek was the proprietor of a nearby forge that had been in existence at least since the early fourteenth century. He was presumably an uncle of Walenty Roździeński. The Brusek family also owned a neighboring forge at Piaseczna. Roździeński speaks in his poem of the town of Tarnowskie Góry; the town records mention an Andrzej Brusek in 1586 and Wojciech Brusek in 1597.[3]

We are somewhat better informed as to the family of Roździeński's mother, whose family tree is given in Table 1. This is a true family of Polish forgemasters; her father was the forgemaster Zych Bogucki, her grandfather the forgemaster Jakub Warchoł(?) from Bilanów, and an uncle, Jan Warchoł, was forgemaster at Przystań and founded a forge at Kostrzyna. Her aunt Małgorzata married a forgemaster at Zawiercie, Jan Marcisz, while another aunt probably also married a forgemaster.

Forgemasters had no guild of their own, as did smiths in towns. In 1547 the feudal lord Jan Kochcicki ordered village smiths in the environs of Lubliniec to join the guild in that town.[4] The guilds in those times imposed certain duties on artisans, but also supported their rights.

The forge at Roździeń, like the neighboring forges at Bogucice, was situated in Silesian territory, separated from the Kingdom of Poland since 1339 when King Kazimierz III gave up his rights to Silesia. This region, divided into small duchies, had previously been subdued by the King of Bohemia, John of Luxembourg.

The Roździeń forge was only a few kilometers from

3. d66:198.
4. c57:135.

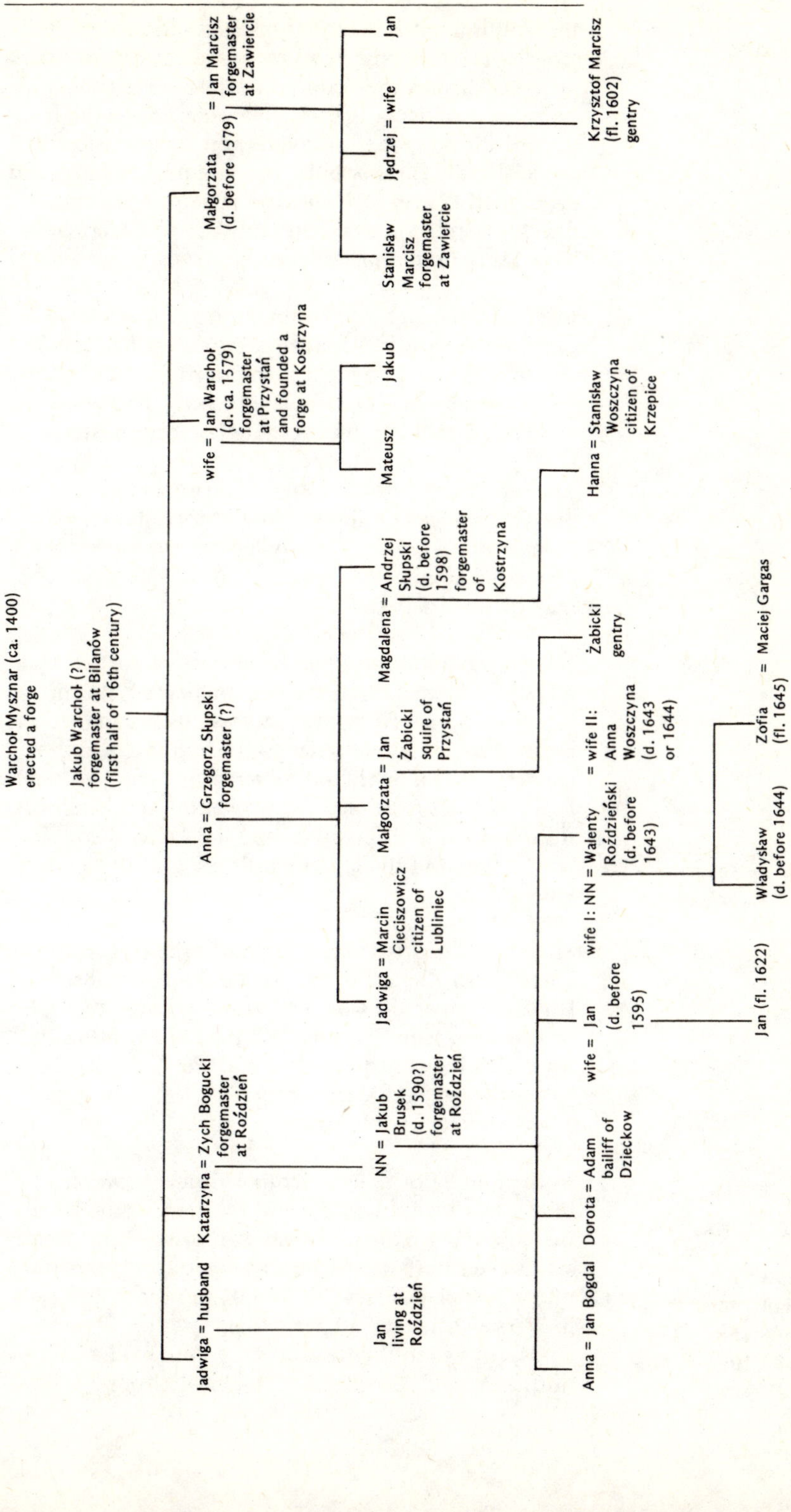

Table 1 Family tree of Walenty Roździeński (from c55:291)

the Polish frontier. As Roździeński's life shows, this frontier did not exist at all for forgemasters--at least as far as kinship and economic affairs were concerned. In Silesia, particularly in regions adjacent to the frontiers of the Kingdom of Poland, the forgemasters and especially the rural population were mostly Poles and spoke Polish, although the official language was Czech. Foreign newcomers–mostly from Germany–were fairly numerous only in the towns.

In 1546 the proprietor of the estate of Mysłowice, Stanisław Salamon, founded the forge at Roździeń and settled Zych Bogucki there, having brought him from the neighboring Bogucka forge.[5] This occurred at a period that was unfavorable for forgemasters.

The second half of the sixteenth century brought important changes in the standing of forgemasters in Poland and Silesia. It is only against the background of these economic changes that the *Officina Ferraria* may be understood.

The time had passed when forgemasters, settled by a feudal lord, enjoyed numerous privileges.[6] In the fourteenth and fifteenth centuries dukes had often granted forge rights similar to those granted to the property of a knight, and forgemasters enjoyed certain privileges reserved only for gentry and landlords; for example, their bequests and their wives' dowries were entered in the land records, as were those of knights.

Such privileges had been granted to the Bogucka forge. When Jurga Kleparski was settled there in 1486 by Duke Kazimierz of Cieszyn, he was granted the right of firewood from the forest of the Mysłowice estates, for which he was to pay an annual rent of only 22 Hungarian red guldens and three cartloads of good iron. This rent could never be raised.[7]

In the middle of the sixteenth century, however, the feudal landlords in Silesia and the gentry who ruled the country in Poland endeavored to seize for themselves ever more numerous privileges at the expense of other social classes. In 1563-1564 the Polish Diet in Warsaw canceled all perpetual leases on Crown lands, for which the forgemasters were paid a sum not to exceed 100 marks.[8] It appears from

5. c28:59.

6. c59:190; c57:121.

7. d66:188.

8. c30:19; c59:196-197.

assessments for taxation purposes, which were certainly reduced, that the value of a forge ranged from 100 to 200 marks.

Landlords could now lease out forges for a year or several years in return for a fixed annual rent. This kind of contract was especially unfavorable for the forgemasters since the landlord could refuse to renew the contract and the forgemaster then had no right to his workshop. Instead of being a free craftsman as before, his status was lowered almost to that of a serf.

The rents paid by forgemasters in the latter half of the sixteenth century varied greatly. Rents were often partly or wholly paid in the form of iron products. For instance, a forgemaster named Klemens from the forge at Ruda near Bzin paid 8 marks yearly and 8 cartloads of *"ferrum bonum"*; forgemasters from Łagów paid 10 marks and 3 cartloads of plough iron.[9] Over and above the rent, the forgemaster was sometimes obliged to supply the landlord with a considerable amount of iron at a very low price (for example, 30 cartloads at half the market price). More frequently, especially on Crown estates, the rent in iron products was replaced by a payment in money. At the Rekszowice forge, founded in 1563, the rent was at first fixed at 8 cartloads of iron and then changed to a payment of 10 złoty and 20 groszy (1 złoty = 30 groszy).[10]

Besides the rent, the forgemaster was taxed, but the taxes were not excessively high. According to a law of 1520, in force in the Kingdom of Poland, the forgemaster paid a tax of 1 złoty for his works and for himself and 6 groszy for his wife and each of his children. The constitution of 1563 fixed the forgemaster's obligations somewhat differently, determining that he should pay 1 złoty for each waterwheel and 6 groszy for each workman: assuming that a forge generally had 3 waterwheels and 6 workmen (although certain forges in Poland employed 10 or even 20 workmen), the total tax amounted to 4 złoty and 6 groszy. According to clauses in the constitutions of 1580, 1581, and 1588 Polish forgemasters then had to pay double this sum.[11]

In addition to the rents and taxes the forgemasters had to pay for the iron ore, or at least for its

9. c2:35.
10. c2:36.
11. c2:40.

transportation, as well as for the workmen who supplied the charcoal—generally derived from wood cut on the estate of the proprietor—and the men working in the forge—the smelter, the smith, the charcoal handler, and other helpers. The smelter received 2.5 groszy for one bloom, the smith got as much for forging one cartload of iron,[12] and the helpers' wages were much lower. It is possible that for unskilled labor, such as might be needed for extracting and transporting ore, the forgemaster relied on socage.

In cases of either lack or excess of water to drive the waterwheels, the costs exceeded the income and the forgemaster had to borrow money to pay his rent. Meteorological conditions and the irregularity of supply of raw materials were also responsible for the bloomery furnace operating for only a part of the working year.

In 1563, by order of Archduke Ferdynand, the Silesian Chamber required forgemasters to declare their annual production. They answered that this was very difficult to determine, since "there is more water one year than another, since better ore is obtained at one time than at another, and, moreover, since one's luck depends upon the availability of good workmen."[13]

Attempts to find information in archival materials concerning the yield of the Roździeń forge in the late sixteenth century have so far been unsuccessful. Such information is preserved concerning the neighboring Bogucka forge, but for a later period.[14] It may be assumed that in Roździeński's time the production of these forges did not greatly differ: Agricola, in his work of 1556, gives similar data. In 1702 the bloomery furnace at the Bogucka forge operated 75 days, giving 4 blooms of metal daily, having a total weight amounting to 5 to 5.5 centners (hence the weight of a bloom was 1.25 to 1.5 centners, i.e., 78 to 88 kg). The production of such a bloom required 7 to 8 kible of ore (one cartload) and 3 kible of charcoal; moreover, the forging of a bloom into bars required a further 1 to 2 kible of charcoal.[15]

Hence the yearly yield of a forge in the late seventeenth century was up to 400 centners, sometimes more.[16] This would correspond to about 350-550

12. c59:140.

13. c60:124.

14. c28:185.

15. The charcoal data clearly require correction in terms of the unit of measurement: the "kibel" (pl. "kible") should be replaced by the "basquet," a unit several times larger. In the case of the bloomery process, the weight of charcoal consumed was approximately equal to that of the ore smelted.

16. c59:79; d56:288.

cartloads of iron, a cartload usually meaning 24 bars. (In Krakow in 1628 a centner of Silesian iron was 18, 20, or 22 bars; evidently they differed in size.)[17]

In the late sixteenth century a bar of iron cost 3 groszy in the Kingdom of Poland, and a cartload of iron about 63 groszy. Thus the value of a year's production (350-550 cartloads) amounted to 758-1191 złoty and 20 groszy.[18] Part of this iron was set aside for paying the rent and taxes, part had to be supplied at a low price to feudal landlord; besides this the forge also required a certain quantity of iron for its own purposes.

The price of iron was subject to fluctuation, especially in Poland, where it rose in the period between 1530 and 1590 by about 60%; for comparison, the price of food rose in this period by about 250%.[19]

At the end of the sixteenth century a devaluation of coins occurred. Before 1577 the value of metal in a Polish grosz remained constant at 0.77 gram of silver, but it was debased in the late sixteenth century to 0.66 gram, and thereafter the value of coins continued to decrease.[20]

The profit a landlord obtained from his forge was restricted to the rent and, possibly, to a certain amount of iron free of charge or at a low price; this was a fairly modest profit, especially since the forgemaster generally used large quantities of wood from the landlord's forests. The opinion was therefore often expressed in Poland that a forgemaster brought in a small rent and "causes great damage to forests and spoils the peasants' bee trees by which the bees are lost and the revenue (therefrom) also." Polish landlords made a much greater profit from developing farmland which produced rye, wheat, and other agricultural products. These were mainly exported via Gdansk to West European countries. The annual profit a landlord got from a farm averaged 344 złoty.[21]

These conditions induced the Polish gentry to close down forges and convert their lands to farming. The waterwheel was sometimes adapted to operate a cornmill. Thus in 1588 the landlord Padniewski closed down the Nicowska forge near Zawiercie, which

17. c53:19.

18. c47 (vol. 2):291; c59:106.

19. c59:166.

20. c46 (vol. 1):210.

21. c46 (vol. 1):152.

supplied royal ordnance materials, and stopped work in the neighboring Marciszowska and Zawadzka forges as well.[22]

The Polish gentry would even use violence in dealing with forgemasters who put up any resistance and tried to defend their work places. In 1605 Zaremba, the *starosta* (sheriff-nobleman) of Grabow, attacked and destroyed the forge of Paweł Rostoga, who came under his authority, and in 1612 he made a similar assault on the forge at Głuszyn, where he beat the proprietress.[23]

The above-mentioned limitations placed on the status of free forgemasters in the Kingdom of Poland took place chiefly in the Crown lands (including estates rented to the nobility) and in the estates belonging to the nobility, during the reign of King Zygmunt III Vasa (1587-1632). In the large landholdings belonging to the clergy in Poland, and especially in the vast estates of the Krakow episcopate (including important ironworking regions in Little Poland), the situation of forgemasters was much better. Nevertheless they experienced some restrictions because their perpetual leases were universally changed into temporary ones and usually into annuity leases. Besides this, sometimes the law allowing the repurchase of forges or the fact that the majority of the adjoining leaseholds belonged to the nobility was very harmful to them. However, the privilege of running such a forge was generally given to a forgemaster who was a specialist in his art, and his rights and activities were in principle protected and supported by the administration of the episcopate estates.[24] It can be said, then, that these enterprises became the mainstay of free forgemasters in Poland.

In those very estates of the Krakow episcopate, in the neighborhood of the free forges at the beginning of the seventeenth century, a new chapter in the history of Polish iron metallurgy was opened. Thanks to the privileges of the year 1610 granted by the bishop, two Italians from Bergamo, Laurence Caccia and his brother John Andrew Caccia, built the first blast furnace in Poland in Bobrza near Kielce.[25]

In Silesia, under the Habsburgs, no such far-reaching changes in the iron-smelting industry occurred.

22. c59:194.

23. c59:201.

24. c59:235; c25:283, 320, 336, 338, 350, etc.

25. c26:827; c27:255.

Nevertheless here, too, the forgemasters' privileges were restricted. When Mikołaj Salamon inherited the Mysłowice estate in 1549, he tried to reduce the privileges of the master at the Bogucka forge. The following year Master Andrzej of that forge brought suit against Salamon in the court at Pszczyna for trespassing and infringing on the privileges of the forge, and he won his case. In 1558 he brought another lawsuit concerning the right to make use of the Mysłowice forests for the needs of the forge, as well as the right to hunt and fish on his lands. Mikołaj Salamon laid claim to these rights as proprietor of the estate. The court, however, took the part of the forgemaster.[26]

The forgemaster Zych, settled at the Roździeń forge, failed to obtain such privileges, and this would have a serious effect on the life of the author of the *Officina Ferraria.* The forge had been set up by Stanisław Salamon on the land of the head of the village, which he had bought, and the rights attached to the forge were strictly limited to the work of the forge. For these rights the forgemaster had to pay an annual rent of 22 Hungarian złoty plus three cartloads of good iron and some agricultural implements.[27]

Master Walenty's father, Jakub Brusek, who initially ran a tavern at Roździeń, took over the forge from Zych, presumably after marrying Zych's daughter. It is believed that Walenty, his brother Jan, and his two sisters Dorota and Anna were born there in the 1560s[28] or 1570s.[29] Walenty presumably acquired his education at one of the neighboring schools, perhaps at Mysłowice.

Jakub Brusek was not lucky in running the forge, and he fell into considerable debt. When Walenty and his brother Jan took over the forge after their father's death, they tried to save themselves from responsibility for these debts, maintaining that the forge was their mother's property and that their father had no right to encumber their mother's dowry. Walenty also tried to enter his wife's dowry in the district register, but his claim was rejected.

In 1591 the proprietress of the Mysłowice estate, Katarzyna Salamon, summoned Walenty Roździeński before the court at Pszczyna and accused him of exceeding the rights bestowed on him. In 1594 she had

26. c28:62.
27. c28:59.
28. a1948:XIX.
29. d8:160.

the forge assessed and ejected the Roździeński brothers, paying them the determined sum, after having deducted the debts contracted by their father.[30]

Master Walenty did not accept this verdict, and he appealed. The verdict passed by the court at Pszczyna on June 28, 1596, went against the forgemaster. He was even confined in the local prison and was to be delivered over to Katarzyna Salamon.[31]

A new forgemaster, Zygmunt, was settled at the Roździeń forge, but he received notice in 1598.[32] In 1602 Katarzyna Salamon entered into a new lease contract with Marcin Niwka. In 1612-1622 Walenty's nephew, Jan, was tenant.[33]

We have no exact knowledge as to the further fate of the author of the *Officina Ferraria.* It is assumed that he was helped in his misfortunes by a Silesian feudal lord Andrzej Kochcicki, proprietor of Lubliniec and of forges in the environs of Koszęcin, who defended Walenty against Katarzyna Salamon and facilitated his going abroad.[34] In any case, we know from court records that Roździeński went to Poland for a certain time, as he had already intended to do in 1595. During that period he was presumably in charge of the Niwka forges, which he mentions in his poem.

Apparently, Master Walenty did not manage these forges for long. An entry found in the archives of the Metropolitan Curia in Krakow mentions that, in 1602, "the noble Walenty Roździeński, having no fixed dwelling, lived for the most part with a certain Walenty Słotowski, of whom nothing further is known."[35] This Słotowski, presumably a citizen of the duchy of Siewierz, acted as Roździeński's plenipotentary in the Episcopal Court in Krakow in a suit involving one Jan Dębski, the details of which are not known.

The poem *Officina Ferraria* was probably written between 1599 and 1607.[36] The author mentions that some of the Niwka forges "were abandoned more than ten years ago." Archival sources show that these forges were closed down about the year 1588.

It is generally agreed that during this period Roździeński was again living in Silesia and that he wrote his

30. c28:66-68; d66: 185.
31. c28:66.
32. c56:94.
33. d22:22-25.
34. a1936:XX.
35. c25:307.
36. d39:33.

poem in the Lubliniec mansion of Andrzej Kochcicki, a well-known patron of science in Silesia.[37] There Master Walenty had access to this feudal lord's famous library. A copy of Schwenckfeldt's work *Stirpium et Fossilium Silesiae Catalogus* (Leipzig, 1600) coming from this library has recently been found. It was probably from this very copy that Roździeński drew his information concerning the occurrence of ores and precious stones in Silesia.

When Roździeński's poem was printed in 1612 (at the press of Szymon Kempini in Krakow), he dedicated it to Andrzej Kochcicki, probably in gratitude for his hospitality at Lubliniec, as well as for his help in the lawsuit against Katarzyna Salamon.

It is astonishing that no mention has been found anywhere of the existence of the *Officina Ferraria* prior to Professor Pollak's publication in 1929, and that only a single copy of it has so far turned up.

The view that the rest of the edition was lost because of the popularity of the work among forge workers has, probably correctly, been criticized.[38] The absence of other copies has also been attributed to Roździeński's marked leaning toward fraternizing with common folk and his criticism of the gentry ("today's evil people who adopt fine titles while slandering the good people they despise by calling them by disgusting names"). Copies of the book might therefore, it is suggested, have been destroyed by order of the gentry.

One more explanation might be noted. Andrzej Kochcicki, to whom Roździeński dedicated his work, was a well-known leader of the Protestant movement in Silesia. In 1620 he headed the delegation sent by the Silesian states to the Warsaw Diet in an effort to win the goodwill of Polish political leaders for Silesian Protestantism.[39] In his dedication Roździeński expresses a favorable opinion of the reformed religion. Moreover, the introductory verse to the *Officina Ferraria* was written by Daniel Murovius, a pastor from Koszęcin.

Meanwhile, the Counter-Reformation had won out in the Kingdom of Poland, and any show of sympathy with dissenters met with the violent reprobation of

37. a1948:XXI; d7:489; d12.

38. d36:594.

39. d7:487; d66:194.

the Catholic clergy, whose strong influence led to the passage of a series of laws against the dissidents. This may be the reason why the copies of the *Officina Ferraria,* published in a rather limited edition, were destroyed.[40]

This is not the only problem awaiting solution. It is not really certain that Roździeński in fact went to Kochcicki's manion at Lubliniec in 1602 and there wrote his work. A mention has been found in archives that Zygmunt Roździeński married Anna in 1617 in St. Mary's Church in Krakow,[41] and that Walenty's son, Władysław Roździeński, was enrolled in 1626 at the Jagiellonian University in Krakow.[42] There is also information about a Jan Roździński (Roździeński?), an assistant scribe at the Episcopal Church at Siewierz, who often came to Krakow; his mother lived there in 1633 and owned a small house. He was 26 years of age and hence was born about 1607.[43] It is not quite certain whether Jan Roździński was a son of the author of the *Officina Ferraria.*

Reading the poem, one feels clearly that Roździeński wrote as a Pole who felt at home in the Kingdom of Poland.[44] He clearly considers the Germans to be newcomers to Silesia and to constitute a foreign element. In describing iron mines exploited at the Niwka forges and those in the environs of Ogrodzieniec, Olsztyn, and Krzepice, Roździeński does not think it necessary to mention that they are located in the territory of the Kingdom of Poland. Yet he does make explicit identifications when describing ores occurring outside the confines of Poland, namely in the Kingdom of Bohemia and in Silesia. Last but not least, such a phrase as "in our country . . . at Olkusz" would scarcely be used in a work written beyond the frontiers of Poland.[45]

Thus both the place where the poem *Officina Ferraria* was written and the poet's biography warrant further investigation by archival researchers.

Some information about Roździeński's later life has recently come from notes found in the archives of the Metropolitan Curia in Krakow. These give evidence that in the year 1644 Zofia, the daughter of Walenty Roździeński, appeared before the court of bishops in Krakow in a lawsuit with Maciej Wróblowicz, the

40. d39:36.
41. c52:116.
42. c3:10n.
43. c54:457.
44. d13:95; d39:34.
45. Olkusz was a town in the Kingdom of Poland that was famous for its production of lead. Georgius Agricola mentions it in his *De Re Metallica* (Basel, 1556).

parson of the small town of Koziegłowy. The object of the suit was the fortune of Walenty Roździeński, consisting of "many things of various kinds, both personal property and estates, cash, jewels, goldwares, silver-, copper-, and tinwares as well as brasses, robes, and clothes, private armor of all kinds, weapons, harquebuses, books, libraries, cultivated ploughlands, and other things valued overall at 2000 Polish marks *[grzywna]*."

As the succeeding heirs of the Roździeński family, Anna Woszczyna, Master Walenty's second wife, and Władysław Roździeński, Walenty's son, were mentioned, but the latter died before the year 1644.[46]

These new sources suggest that in the last years of his life Walenty Roździeński and his family were connected with Koziegłowy in the Polish Kingdom, and may even have lived in this little town. The parish church (Figure 3) has on its doors some fine wrought-iron hinge straps (Figure 4) which nicely illustrate one of the more important end uses of Roździeński's labor.

To the reader what is of importance is that Roździeński has bequeathed us in his poem a true and unfeigned cult of labor, a thorough knowledge and love of the craft he practiced and of liberty, which are nowhere better expressed than in the words of his own last chapter, "The proper model or image of the life of a smith":

"So don't ask in our smelting house about the wealth that others long for in the world; you see how painfully we labor and yet we suffer want and misery.

"Although in misery, we nevertheless enjoy the liberty which we have from our forefathers. It is our liberty alone in which we rejoice, as this is the greatest reward for our misery—not treasure and not money!"

II. The *Officina Ferraria* and Its Place in Poetry and the Literature of Metallurgy

Walenty Roździeński's poem *Officina Ferraria or the Smelting House and Forges of the Noble Craft of Ironwork* was published in 1612 "at the printing office of Szymon Kempini" in Krakow. The book consists of 46 leaves, 18 X 13 cm in size, and is

46. d29.

Figure 3 Parish church at Koziegłowy, built in the 15th-17th centuries. The roof is covered with wooden shingles. Photograph by S. Miczulski.

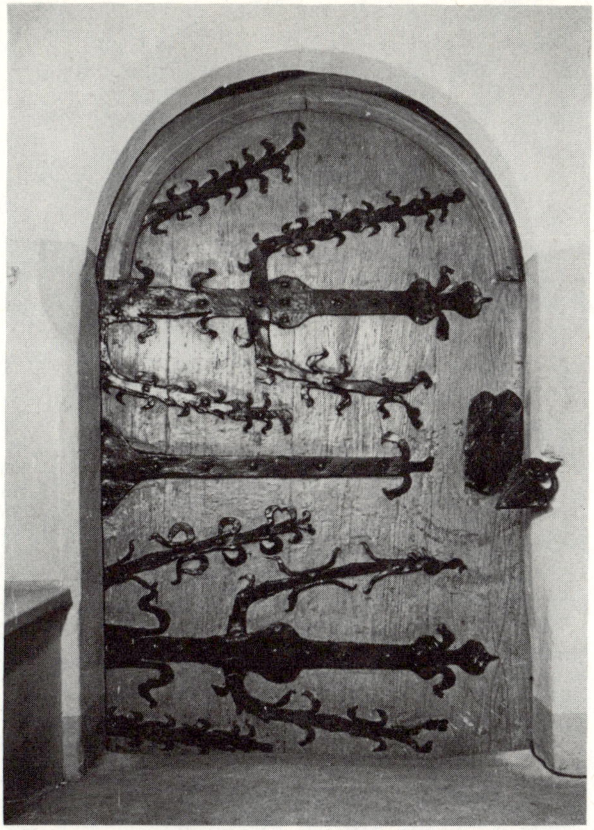

Figure 4 The wooden door of the church at Koziegłowy with iron furnishings dating from the 17th century. Photograph by S. Miczulski.

illustrated with six woodcuts. The leaves are unpaginated but bear signature marks A, A2, A3, [A4], B, B2, . . . , M2. (The beginning of each leaf and its verso is noted in the translation.)

The work consists of two quite different parts. The first contains the history, as it was then understood, of the discovery and spread of the art of smelting and ironwork in Europe, as well as information on the occurrence of nonferrous metal ores and precious stones in Silesia. This part is compiled from several sources and goes to prove the erudition of this simple forgemaster. The second part, beginning with a description of the history of iron in Silesia, is essentially a record of the author's personal experience and practice in metallurgy, as well as of his own life and deliberations.

The work starts with two brief epigrams, written in Latin, by Daniel Murovius, pastor at Koszęcin and Turnawa, and Pawel Twardock, principal of the school at Lubliniec.

Roździeński then offers a laudatory verse in honor of the arms of the house of Kochcicki, the "Heron." A similar poem was published in 1612 by F. Taubman, who in numerous verses extolled people of renown living in Silesia.[47] Although the themes of the verses are identical, there is no evidence that either author used a previously written work as his model.

Next, Roździeński formally dedicates his poem to Andrzej Kochcicki, a Silesian feudal lord and proprietor of forges. This dedication is similar in form to the epistles customary in those days, but it differs in its lack of superfluous rhetoric and exaggerated eulogy, a fact which bears witness to the simplicity of the author and also to his personal dignity.

In his dedicatory address Roździeński gives an outline of the contents of the poem and proclaims the virtues of the Lord of Lubliniec, especially his religious attitude at a period of "confusion of creed." This fragment proves that Master Walenty was a Protestant, since Kochcicki was one of the most active adherents of the Reformed Church in Silesia.

The compiled part of the *Officina Ferraria* begins with a verse in which Roździeński cites the Bible to

47. b8.

introduce the reader to Tubal-cain and Enoch, "who were the first smiths in the world to discover iron and copper and the different crafts by which they are made serviceable by treatment with fire." He then writes of Hiram, king of Tyre, and of the ironworks in the land of Cabul, and relates the legends of Vulcan and of the Cyclopes. Along with information derived from the collection of sermons to miners by J. Mathesius, *Sarepta oder Bergpostill* (first published in Nürnberg in 1562 and repeatedly reprinted), Master Walenty takes the opportunity to mention bog iron ore, well known to him from his own practice.

Roździeński goes on to present the history of the spread of iron in Europe and southwestern Poland. First he goes back to the ancient writers, when writing "On certain smiths to whom the discovery of iron in the second age of the world [i.e., after the deluge] is ascribed" (see Figure 5). The Idaedactyli were said to be the inventors. Walenty cites Pliny and Strabo on the Cyclopes and the Chalybes, as well as the Telchines from Rhodes. Nevertheless, he believes that the first forgemaster was Tubal, descendant of Noah, who came to Spain and there built iron-smelting works. His knowledge passed to his descendants, the Carthaginians and the Chalybes; the latter were said to have invented the process of quenching steel.

Citing Justinus, Roździeński also gives information on the special properties of the water of the rivers Bilbilis and Chalybs in northern Spain, thanks to which the tools quenched there were extremely sharp. He considers that this information is confirmed by the high quality of sewing needles produced in Spain.

The part of this section concerned with fable concludes with information about Cadmus of Phoenicia, who was said to have found galmei (cadmia, calamine) near Thebes and to have obtained brass therefrom.

The further history of iron in Europe as described by Roździeński was presumably based on tradition; in any case it does not differ much from the views held at the present day as shown in Figure 6. Thus knowledge of iron smelting was brought from Greece first to Slovakia, Serbia, Bosnia, and Hungary and then to Carinthia and Styria, where the technology of iron smelting was conveyed by Slovak smelters. This

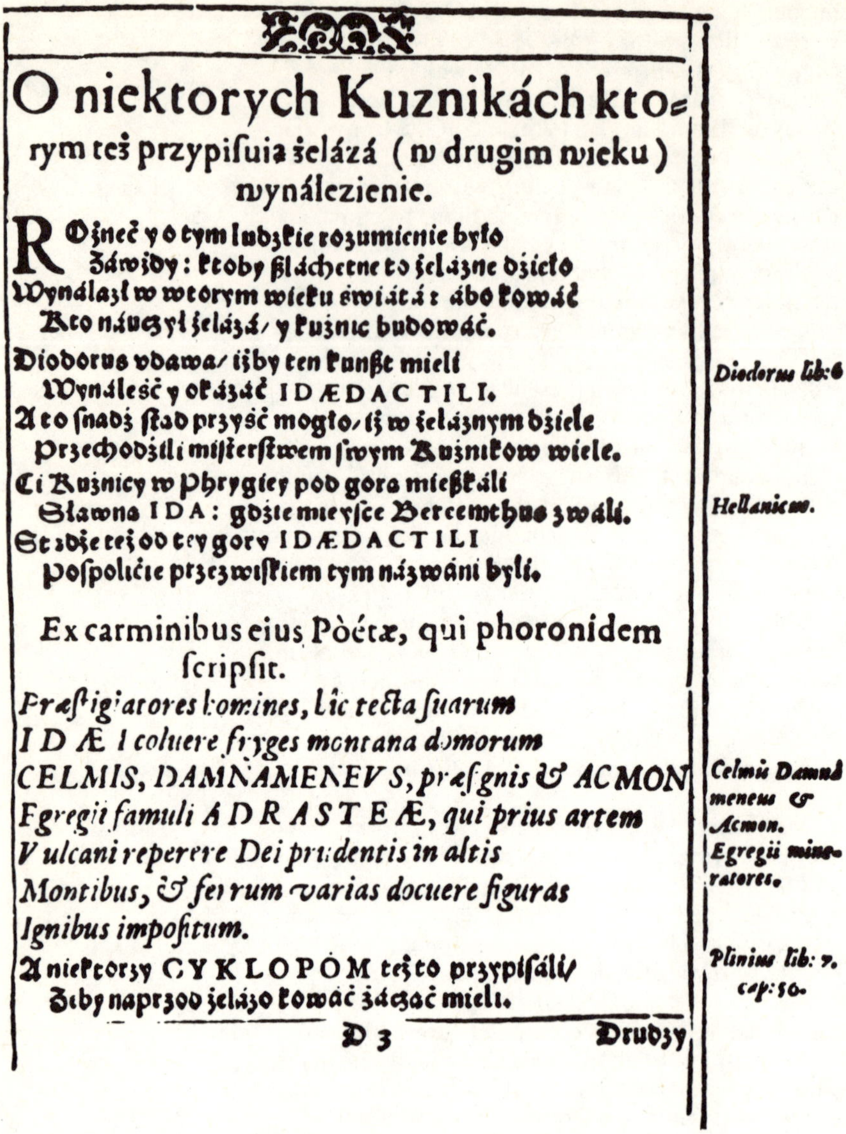

O niektorych Kuznikách ktorym też przypisuia żelázá (w drugim wieku) wynálezienie.

ROżneć y o tym ludzkie rozumienie było
Záwżdy: ktoby ślachetne to żelazne dzieło
Wynálazł w ktorym wieku świátá: ábo kowáć
Kto náuczył żelázá, y kuźnie budowáć.

Diodorus vdawa, iżby ten kunszt mieli
Wynáleść y okázáć IDÆDACTILI. — *Diodorus lib: 6*
A to snadź stąd przyść mogło, iż w żeláznym dziele
Przechodźili misterstwem swym Kuźnikow wiele.
Ci Kuźnicy w Phrygiey pod gorą mieszkáli
Sławną IDA: gdźie mieysce Berecemthus zwáli. — *Hellanicus.*
Stądże też od tey gory IDÆDACTILI
Pospolicie przezwiskiem tym náżwáni byli.

Ex carminibus eius Pöetæ, qui phoronidem scripsit.

Præstigiatores homines, hic tecta suarum
IDÆ l coluere Fryges montana domorum
CELMIS, DAMNAMENEVS, præsgnis & ACMON — *Celmis Damnameneus & Acmon.*
Egregii famuli ADRASTEÆ, qui prius artem
Vulcani reperere Dei prudentis in altis — *Egregii mineratores.*
Montibus, & ferrum varias docuere figuras
Ignibus impositum.

A niektorzy CYKLOPOM też to przypisáli, — *Plinius lib: 7. cap: 56.*
Iżby naprzod żelázo kowáć záczać mieli.

D 3 Drudzy

Figure 5 Page D3 of the 1612 edition of the *Officina Ferraria.*

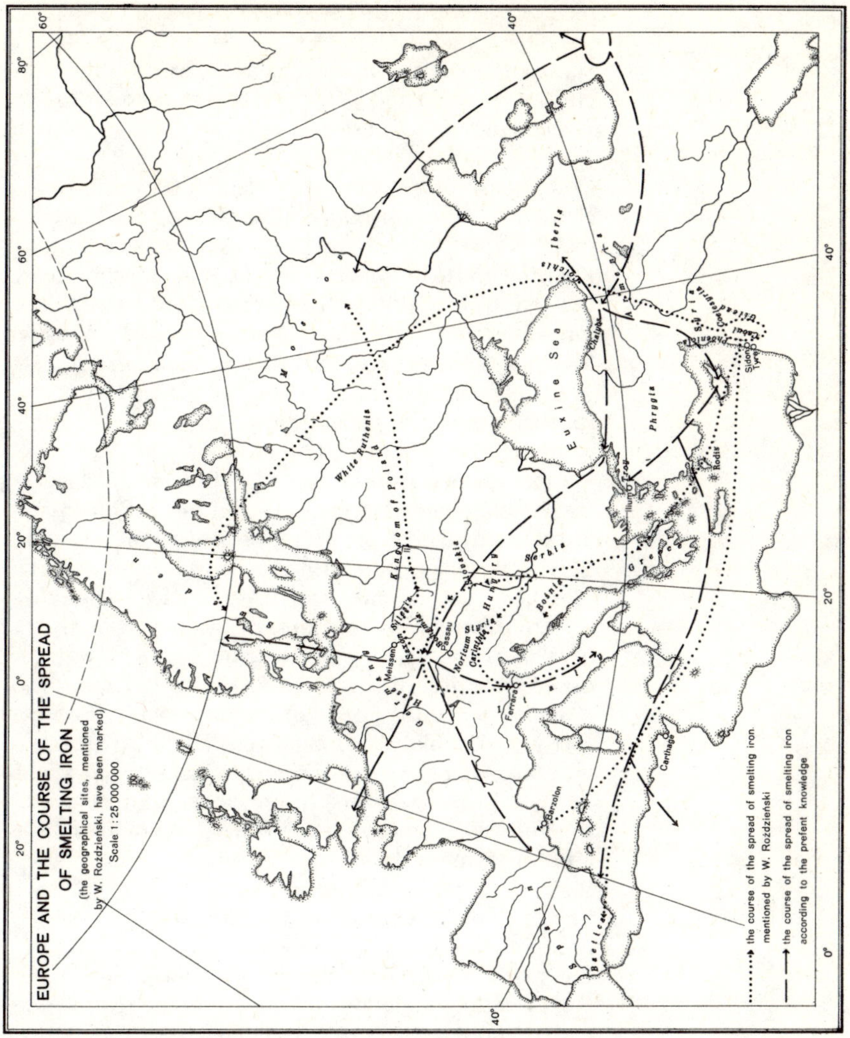

Figure 6 Map of Europe showing the course of the spread of the art of ironworking.

technology was further spread to Italy and Germany, the oldest smelting houses there being those at Noricum and at Passau. In Roździeński's opinion the craft of smelting and working iron was brought to Sweden direct from Colchis.

On the basis of a statement in Haiek's Czech chronicle of 1541[48] (omitted from the German translation of 1596), Roździeński goes on to say that the first smelting house in Bohemia was built at Zdechovice in the year 677 by Botak. This presumably refers to the introduction of an elevated shaft furnace, the so-called low bloomery furnace, for iron had been smelted in Bohemian territory in hearths or molded pit furnaces for many centuries previously.

Master Walenty also states that, in his time, high furnaces were being used for iron smelting in Bohemia. These were high bloomery shaft furnaces, also called Styrian furnaces, an intermediate form between a low bloomery furnace and a true blast furnace. The Polish smith criticizes the course of the process in the Styrian furnace, saying that it takes a long time and uses a great deal of charcoal. The latter objection was also brought to the notice of the Chamber by Bohuslav Jáchym of Lobkovice in the Czech Rudavy (Krušnohoři Mountains) when Zacharias München built a blast furnace for iron, the first in Bohemia, at Šmiedeberk; this furnace competed with the neighboring bloomeries of Jáchym.[49]

The technology of smelting is affirmed to have been brought from Bohemia to the province of Meissen in the year 974, and from there carried on to Silesia. Roździeński enumerates the forges on the river Małpadew and the works near Lubliniec belonging to Andrzej Kochcicki. He says, moreover, that in Silesia the Poles acquired their knowledge of ironworking from Germans who came from Meissen; however, it follows from what he says later that this learning concerned only the use of water power.

According to historical sources, knowledge of the use of water power in the metallurgy of iron came from western Europe, reaching Meissen first; it was known in Bohemia in the fourteenth century.[50] Mention of waterwheels driving bellows and hammers in Polish forges also dates from this period. Though this

48. c17.
49. c21:80.
50. c22:24.

improvement may have been introduced in Silesia by newcomers from Meissen, in Little Poland (the southern part of the Polish Kingdom) it was introduced by monks of the Cistercian order, who came from the maternal monastery at Cîteaux in France. The Cistercian monasteries at Jędrzejów and Wąchock—founded in Little Poland in 1149 and 1179, respectively—may have played a role. This may be assumed from the privilege granted by Bolesław Wstydliwy, Prince of Krakow, to the monastery at Jędrzejów in 1261.[51] Roździeński states that the knowledge of water drive was later brought further eastward by Polish forgemasters into the land of Ruthenia, and this is confirmed by archival sources.[52]

In concluding the history of iron in Silesia, Roździeński mentions his ancestors, Hercyg and the Bruseks, as well as his invented coat of arms, the "Serdecznik," to which he devotes a short verse.

Having finished this historical introduction, Master Walenty proceeds to an original description of the forges in Silesia that were known to him personally (Figure 2). He describes the beginnings of the Bogucka forge and the forgemaster Jurga Kleparski who had settled there; next he enumerates the forges in Little Poland. These were, of course, not all the forges existing in Silesia and western Little Poland at Roździeński's time. This section of the poem ends with a verse devoted to the work of the miner.

The section following, entitled "On iron ore, which is found in certain countries in various shapes," is unquestionably one of the most valuable and original parts of the *Officina Ferraria.* The author describes in detail the iron ore deposits occurring in Little Poland and the environs of Bytom in Silesia; the description of other Silesian ore deposits, as well as of the ore exploited in Moravia, "in the domain of Szternberg," is, however, less detailed.

Here Master Walenty relies entirely on his personal experience. His observations on the procedure of smelting particular iron ores and on the mixing of various kinds of ore to obtain a metal of the best properties are most valuable and original. Except for Biringuccio's work no such information is to be found in the better-known metallurgical treatises of the period.[53]

51. c59:89.
52. c50:60; c9:193; also c59:205; d56:234.
53. b7; see d39:48.

From information scattered throughout the poem, especially toward the end, it is possible to reconstruct the metallurgical iron-smelting process in use in Silesia and eastern Little Poland. It was carried out in a low bloomery furnace with a round pit. In the case of Silesia, this is confirmed in a note by J. Gesner to the German translation of Nicolas Monardes's *Dialogo del Hierro,* published in 1615.[54] Such a bloomery furnace was described in detail by Georgius Agricola in his famous work *De Re Metallica* in 1556.[55] The size of the hearth varied somewhat, depending on the yield of the ore; it was generally about 42 cm in diameter with a depth of about 28 cm.

After the furnace hearth had been properly preheated, it was charged alternately with charcoal and with ore, which in certain cases had been previously washed and/or roasted; the fire was urged with a stream of air directed from the bellows through the tuyeres. After a few hours of smelting, an irregular, spongy lump of reduced metal mixed with slag formed at the bottom of the pit. This lump, called a bloom, was extracted and hammered with a wooden hammer to free it of the cinder sticking to its surface; next it was cut into several pieces, which were shingled to remove most of the slag inside and forged into bars.

The end result of this process was mild bloomery iron with a purely ferritic structure, containing a certain amount of slag in the form of inclusions. It was also possible to obtain a steel bloom by changing the angle of inclination of the tuyere.[56] The possibilities were limited, however, and depended upon the chemical composition of the ore, especially on the phosphorus content.[57]

When smelting a low-phosphorus ore—ore with less than, say, 0.75% P_2O_5—ancient smelters managed to achieve the carburization of a large part, or even the whole, of the bloom and thus obtained steel during the smelting operations. On the other hand, when ore with a high content of phosphorus was smelted—and Roździeński had mostly to deal with such ore in practice—carburization was attained only on a small part of the bloom, on the exterior of its lower surface. These carburized parts are called by Master Walenty *grąpie* (singular *grąp,* pronounced "gromp"). They were separated from the bloom by quenching in water

54. b12.
55. b4; see c34:179.
56. c13:40.
57. c35:378.

and hammering and were then wrought and forged together; the product of this process (described by Fullon in 1819)[58] was *dul,* a form of steel with a phosphorus content of 0.08-0.40%.[59] Formerly it has been suggested in the literature that *dul* was a low-carbon steel.[60]

Thus the forgemasters in Little Poland and in Silesia produced, besides wrought iron, *dul,* a variety of steel of inferior grade. Several times Roździeński mentions *stal* (steel)—also a carburized metal, but containing less than 0.08% phosphorus—but only as a metal produced abroad (for example, in Styria).

As the phosphorus content of the ore was variable, there was no distinct borderline between the *dul* and steel. Thus a *dul* of the highest grade, containing probably 0.08-0.15% phosphorus, such as was made in the environs of Bytom of Piekary or Żyglin ore, and also in the Niwka forges, was of a quality approaching steel, as Roździeński rightly remarks.

At the end of this part of his work, Master Walenty blames the officials of the Duchy of Bytom, who were mostly Germans, for their negligence which caused losses in the ore mines there.

Here there is a break in the construction of the poem: in the section entitled "On the disappearance of the Bytom ore and the emergence of the Tarnowskie Góry ore," the poet relates a local legend, not found in other sources,[61] about a demon named Szarlej and the vanishing of the silver ore near Bytom. The statement that a peasant named Rybka found ore near Tarnowskie Góry seems likely, while the information that the town of Tarnowskie Góry was founded by Duke Janusz of Opole in 1526 is true and is confirmed by archival sources.

The following three sections—"On the ores and precious stones found in the earth in certain places in the Duchy of Silesia," "Fiery exhalations, or the emissions of fume from ores," and "On ghosts or mountain goblins"—are based entirely (without acknowledgment of the source) on G. Schwenckfeldt's *Stirpium et Fossilium Silesiae Catalogus* (Leipzig, 1600); Roździeński also adopted from this work the

58. c14:6.
59. c36:55; d39:50.
60. d54:LXXIX.
61. d26:247; d60:254.

motto from Job which he placed on the title page of his poem.

In the following section on "Dwarfs or goblins of the forge," Roździeński returns to the main theme of the poem, writing once more of legendary spirits, this time of those that appeared in ironworks. Such a spirit was said to appear at night, near the "small furnace" (i.e., the smith's hearth), in the Kotowska forge. This was situated on the river Rybna, and Walenty mentions it in describing the Silesian forges.

At this point Roździeński abandons legend and compilation from other sources. The original part of the *Officina Ferraria* then follows. As an experienced forgemaster, Roździeński presents now "The proper organization of a forge." He starts with general advice on managing a forge; then follow more detailed instructions concerning the erection of the bloomery and the superposed chimney, the smith's hearth, and the hammer, bellows, and bloomery implements.

Master Walenty goes on to discuss the work of the bloomery-smiths and blacksmiths, warning against their lack of skill, negligence, and thievish habits. Then he gives the functions of the charcoal handler, whose duties were more diverse than his title would suggest. He had to take delivery of the charcoal supplied by the charcoal burners and charge it into the furnace, to control the fire, and to help the bloomery-smith in other ways. He was also responsible for the millpond and the supply of water to the wheels driving the bloomery and hammer bellows.

The author describes in detail the work of the charcoal burners, who cut wood in the forests and charred it in piles. He considers that covering the piles with turf and kindling them from the top is less effective and more labor-consuming than the method used in Germany of cladding the pile with brushwood or grass and kindling it from the bottom.

In his next section, "On iron, which surpasses every metal in the world by the power of its virtue and excellence," Master Walenty praises iron, making use, in the fashion of those times, of allegories from Greek mythology and adding as an illustration a woodcut of Perseus, who beheaded Medusa with an iron sword.

The conclusion of the poem, "The proper model of the life of a smith," is quite moving. Here the author describes the hard and dangerous toil in a bloomery—and its satisfactions—in most expressive terms, not to be found in other works of the kind.

The *Officina Ferraria* is written in correct Polish, though it contains traces of pronunciation characteristic of the Silesian dialect.[62] The poem contains insertions in Latin, drawn mainly from classical authors. The only Greek word has been spelled incorrectly (*ἡφαὶσον* for Ἡφαίστος). On the other hand, the work does not contain any German words that have not been modified by local usage.

The description of the technology connected with the production and working of iron is especially rich and interesting. Sometimes notions are to be met with which have long been out of use and are not to be found in dictionaries. Hence they are, in certain cases, difficult to explain.

Besides many purely Polish names—for instance, *ruda* (ore: in particular, iron ore), *kruszec* (either nonferrous metal ore or metal in general), *piec dymarski* (bloomery hearth), *żelazo* (iron), *żużel* (scoria, slag, or cinder), *grąpie* (detached fragments of a carburized iron bloom; the word has no equivalent in any language other than Polish), *młot* (hammer), *klyszcze* (tongs), *nakowalno* (anvil), etc.—there are words of Czech origin which occur infrequently—*strój* (instrument, implement), *nastroj* (gear, machinery)—and also words of Latin origin—*forma* (tuyere), *fodyna* (mine, pit). On the other hand, words derived from the German are numerous, especially those which concern hammers—*bana* (hammer face), *buksze* (bushes), *ryttel* (recoil beam)—or the names of the smith's implements—*hesprys* (a hot set), *zynkiesz* (piercer), *stâchel* (crowbar), *formyzen* (tuyere-hook), *szrotyzn* (shears).[63]

It would be wrong to assume that the adoption of many technical terms of German origin proves that the Poles necessarily had acquired all their knowledge of particular technological processes from the Germans, even in Silesian territory.

For example, the following names were adopted from

62. d65:357; d20:166.

63. d40:526; a list of translations used for metallurgical terms may be found on pp. 114-115.

the German language and are still in current use: *stal* (steel, from the German *stahl*; Slav peoples used *ocel*); *hartować* (to quench, from the German *härten*; the ancient Polish term was *kalić*); *cementować* (to carburize, from the German *zementieren*). It appears from investigations of ancient iron objects that Polish metallurgists obtained steel, applying heat treatment and the carburization of iron, as early as the sixth century of our era,[64] that is, at a period when these processes were very probably still unknown to tribes of eastern Germany.

Roździeński uses the original Polish geographical names when referring to Upper Silesia and Opole, but when speaking of Lower Silesia he uses German names in a polonized form, without giving the Polish names that were still in use there.[65] Some of the German names are obviously distorted, such as "Hayelen," probably used instead of "in Heiden" as given by Schwenckfeldt.[66] These latter names were presumably taken from the words of the Silesian-German scholars Roździeński had read.

As poetry, the *Officina Ferraria,* except perhaps the last part of it, yields in excellence to the works of the most prominent Polish poets of the sixteenth century, such as Jan Kochanowski, whose work Roździeński is said to have taken as a model. A proof of this is the great number of pairs and rhymes to be found in both the *Officina Ferraria* and the works of Kochanowski.[67]

The number of ancient sources and works of humanist authors cited by Roździeński in the eruditional part of his poem is quite remarkable. Yet the two sources from which he drew most frequently are not mentioned. These are Mathesius's *Sarepta oder Bergpostill* and Schwenckfeldt's *Stirpium et Fossilium Silesiae Catalogus.*[68] For example, a fragment of Roździeński's poem about the exploitation of copper ore on the river Bober can be compared with the original description by Schwenckfeldt in the following translations:

Schwenckfeldt	**Roździeński**
The Copper Hill, commonly called Kupferberg, a little old town of forgemasters, situated among	These forges were located near the old borough of Kupferberg on the bank of the river Bober; they were

64. c37:19.

65. d61:335.

66. b15:378.

67. d58:XXXII; d59:73.

68. The two sources were identified by E. Haertel; see d3 and d9. More detailed comparison is given by J. Łanowski; see d25:152.

the mountains, on the river Bober, very famous in the bygone days for extremely rich copper veins. . . . They say there were over seventy smelting houses in which copper was smelted in the valley, which in turn got the name Schlackengrund because of the huge masses of slag, most of which were thrown into the river Bober. But only one of the smelting houses has remained till the present day. . .

properly constructed in a very brief time, to the number of seventy. Now they are all abandoned. The only marks remaining on this place are cinders and hammer scale; hence the place where this copper was forged is called Szlakengrunt on account of the slag found there. Of these seventy forges on the river Bober, one workshop still remains. . .

The Bible is the book most often quoted by Roździeński: Genesis four times; Deuteronomy twice; also Joshua, 1 Kings, and the Apocalypse (Revelation) of St. John the Divine, the only book quoted from the New Testament, twice each. There is also one quotation from Job and one from the apocryphal Ecclesiasticus. Of these 14 Biblical quotations, 13 are certainly taken secondhand from Mathesius's *Sarepta,* one of the erroneous footnotes in which Roździeński corrected.[69]

In the *Officina Ferraria* we find excerpts and information drawn from many sources, all dutifully cited in marginal notes. Roździeński quotes from the works of the Greek poets, such as the *Iliad* and the *Odyssey* of Homer, Hesiod's *Theogony,* the *Histories* of Herodotus, *The Trojan Women* of Euripides, the *Lavacrum Dianae* of Callimachus, and the *Argonautica* of Apollonius of Rhodes. Strabo is quoted four times, and other authors are also mentioned: Aristotle (strictly speaking the pseudo-Aristotelean work *De Mirabilibus Auscultationibus*), Apollodoros, Diodoros of Sicily, Dioscorides, Galen, Hellanicus, Phylarchus, Plutarch, and Tzetzes.[70]

Roździeński mentions Josephus Flavius, Cicero, and Pliny three times; Virgil is mentioned four times. Justinus, Lucretius, Macrobius, Ovid, Piso, Mela Pomponius, Tacitus, and one Isatius (possibly

69. d25:154.
70. d25:142.

Eustathius of Thessalonica) are also invoked.[71] Berosus, mentioned in the work, is most probably drawn from the counterfeited *Berosi Deflorantio*, a work written and published in the fifteenth century by Giovanni Nanni (Johannes Annius).[72]

Roździeński drew certain facts from Boccaccio's *De Geneologia Deorum*, written about 1370, including references to Albericus and Theodontius. Among the other Renaissance authors quoted are Aventinus (Johannes Turmair) and Henricus Rutin (perhaps Burchard or Brocard).

Presumably Roździeński was acquainted with few of these sources in the original and knew them mainly from compilations and, in the case of the Greek authors, from Latin translations. Efforts to identify all of these works have so far been unsuccessful, as in the case of the *Liber Exemplorum de Primo Praecepto*, presumably belonging to humanistic literature.

Though the *Officina Ferraria* is the first poem on the iron-smelting industry to be written in a living national language, two Latin poems had been published earlier.[73] The first of these, a poem of 351 lines entitled "Ferraria," was published in Paris in 1533 as part of a collection of verses by Nicolas Bourbon (Nicolaus Borbonius).[74] In a Latin hexameter, Bourbon recalls the blast furnace plant at Vendoeuvres, in France, once directed by his father and just destroyed by a fire. The young poet was greatly impressed by the labor of charcoal burners in the forests, set against a background of forest life full of animals and phantoms of the night, by the prism-shaped blast furnace, the foamy stream driving the hammer, and the rough cheerfulness of the workmen on payday. Yet Bourbon's impressions somewhat embellished the true picture of the ironworkers' life, as well as the technological details of the work they did.

Another Latin poem, *Fabrilium Silesiae Officinarum Fodinarumque Descriptio et Denotatio Brevis* by Krzysztof Winter, first published anonymously at Frankfurt on the Oder in 1556,[75] contains 302 lines and is similar in character to Bourbon's work. At the beginning of his poem, Winter gives a fairly detailed description of Silesia between the rivers Nysa and Bober; then with keen sympathy he presents the

71. d31:360.
72. d25:155.
73. d51:64; d39.
74. b8.
75. b19.

ironworkers' toil. To the eyes of the poet the world of these artisans had an exotic and idyllic charm, devoid of all discord; this feeling is especially emphasized in the final part of the poem, describing the ironworkers' revels.

Another literary work that should be compared to the *Officina Ferraria,* though not in rhyme, is the *Dialogo del Hierro y de Sus Grandezas* of the Spanish physician Nicolas Monardes, published in Seville in 1571.[76] This was one of a series of essays, dealing mainly with the medicinal uses of plants from the New World, that are of considerable interest to historians of botany and medicine. The *Dialogo* was translated into Latin and then into German under the title *Ein nützlich und lustig Gespräch von Stahl und Eisen* (Leipzig, 1615). It is interesting that the German edition was dedicated to Andrzej Kochcicki, to whom Roździeński had already dedicated the *Officina Ferraria.*

The *Dialogo* is a compilation from many sources, including the works of the alchemists, and contains information on metals other than iron. Monardes begins by describing briefly the smelting of iron in the Cantabrian Mountains and then, at greater length, the advantages to man of objects made of iron and steel. This popular-scientific treatise, though containing a certain amount of original material, was written by a person who, not being a trained metallurgist, could only pass on acquired information.

These three works must all yield to Roździeński's poem both in authenticity and in the amount of technological information they contain. However, although Master Walenty includes a great deal of technological information in his poem, even supplementing the discussions of iron in the metallurgical treatises of Biringuccio and Agricola, the *Officina Ferraria* can by no means be classed as a treatise of that character. There is, curiously, no indication that Roździeński knew either of these works.

It has been suggested that Roździeński may have taken some information about spirits or mountain goblins from Agricola's *De Animantibus Subterraneis.*[77] But, on the other hand, he did not take a single word from Agricola concerning iron,[78] while making use of

76. b12.
77. d41:120; d46:119. d53:218.
78. d39:45.

the description of ores in Silesia given by Schwenckfeldt.

Roździeński did not need to consult the earlier metallurgical treatises. He knew much better than Biringuccio or Agricola could describe how a bloomery furnace was built and how iron ore was smelted. He knew that different ores do not smelt in the same manner, that they yield metal with differing properties, and that to increase the efficiency of the smelting operation, combinations of certain ores must be used.

The technological information that Roździeński gives in the *Officina Ferraria* is drawn from his own experience, acquired through many years of labor. It is to be regretted that the poem does not contain any precise quantitative technological data, but, of course, such data were never given in works of this kind.

For this reason the poem of the Polish forgemaster Walenty of Roździeń must be treated primarily as a literary work. It is an unique contribution to world literature—a poem by a simple workman to the glorification of his heavy toil. He was not an educated poet coming from mere curiosity to watch the work of a forge and trying to impress the reader with carefully chosen words and verses: he was an ironworker, experienced and conscious of the social significance of the work he performed, who addressed his reader, inviting him into his workshop: "Just step in to see us, and watch how everyone must continually toil with fire in our smelting house! Look how my body is broiling! Is it little pain that we have in this fire?"

It was a reproach to society—or, rather, to the feudal lords and those classes of society which ordered economic life—that Roździeński wrote his words on the grievous economic situation of the ironworkers, whose pay sufficed for little more than food and drink. This may not have been true of the forgemasters managing the ironworks, whose earnings may have been fairly considerable, but it was certainly true of bloomery-smiths, blacksmiths, charcoal handlers, and charcoal burners.

Walenty Roździeński's poem was not written just to interest and amuse the reader. The *Officina Ferraria* is a work of a social character, a profoundly human piece of writing, and in this lies its greatest value.

OFFICINA FERRARIA,

ábo

Hutá y Wárſtát z Kuźniámi ſzla-
chetnego dźieła Zelaznego.

Przez
WALENTEGO ROZDZIENSKIEGO,
Teraz nowo wydána.

Hiobi. 28. ℣. 1. 2. 6.

H*Abet argentum venarum ſuarum principia: & eſt locus auro quod linquant Metallarii: Ferrum è ſtrictu-ris ſeu glebis venaꝗ; Ferraria ſumitur: & lapide æroſo funditur Æs. Terra eius lapides loco Sapphiri: Pulueres verò ſunt illi.*

W KRAKOWIE,
W Drukárni Symoná Kempiniego/
Roku/ 1612.

OFFICINA FERRARIA

or

**The Smelting House and Forges
of the Noble Craft of Ironwork**

by

Walenty Roździeński

Now newly published

Job 28: 1, 2, 6

Surely there is a vein for the silver, and a place for gold where they refine it. Iron is taken out of the earth, and brass is molten out of the stone. The stones of the earth are the place of sapphires and it hath dust of gold.

Krakow
At the Printing Office of Szymon Kempini
1612

Av

On the Officina Ferraria of Master Walenty Roździeński

Who first began to dig out the seeds of copper from the bowels of the earth, and when; and who, having constructed furnaces, smelted iron in strange ways with constantly raging fire; and how this craft, perfected by industry and art, may be of use to man: All these things does this book teach. After having perused it, dear reader, wish the author good fortune. *Vale!*

Daniel Murovius, Silesian,
a pastor of the illustrious
and distinguished Lord
Andrzej Kochcicki, of the
free Barony of Kochcice
and Lubliniec, in Koszęcin
and Turnawa, etc.

A2

Epigram for Master Walenty Roździeński Writing on the Craft of Ironwork

Different things delight different people, and not all poets have cared to write on a single theme. One sings of arms and the man, another of the titles of the powerful. Still another raises to the stars the fame of the mighty, while very many laud famous towns with lofty walls and splendid houses in noble cities. There are also those whose work is devoted solely to praising magnificent nuptials and funerals and adorning with their words the race of heroes.

Your Calliope, Roździeński, leads your efforts to better endeavors and prospers your intentions. When you call to mind he who first began to forge iron and when you teach of its virtues, your renowned skill rises aloft and the wings of genius carry you toward the sun, inasmuch as with the help of the Muses and Apollo you treat of a craft that is both noble and useful to mankind.

Pawel Twardock of
Strzelce, Silesian,
Rector of the School
at Lubliniec.

On the Ancient Arms of the Honorable Kochcicki Family A2v

The two herons here depicted were granted as arms to the ancestor of the honorable house of Kochcicki, whose members enjoy great fame owing to their virtuous deeds. One of the herons is placed low on a dark blue field, and at the top is another heron. Both stand firmly on logs and hold in their beaks small fish, their own prey, which symbolize the delightfulness of life when one provides one's own food.

The two herons are separated by a helmet and a crown; this was done on account of the high worthiness of the family; this means that everyone, tested as to his virtues, may deserve a royal crown. A3

Herons, as is truly told by those who practice fowling, have peculiar ways of preserving their lives, ensuing from their nature and differing from those of other fowls. They are very quiet birds, unwilling to cause harm as other birds are wont to do; they remain always near water, catching small fish and feeding on them. Though of a quiet nature, the bird must nevertheless, on account of its young, wage war at times with other hostile birds of prey. When the moment of fighting comes, the heron applies clever stratagems. If it cannot succeed by force, it deceives its foes by artifice. One such stratagem, if a falcon attacks from above, is to take the measure of its swoop; the heron then lies in ambush and holds up its beak like a spear against the swoop. If the bird of prey strikes hard, it will harm itself more than the heron, which remains alive while the falcon dies upon its adversary's beak.

If conditions seem threatening, the heron retires into the vast forests when hatching time arrives in spring; there it strongly builds its nest at the top of a high tree, where it brings up its young.

A3v It has such foresight that when it feels a violent tempest coming, it quickly rises high up above the clouds and is thus safe from any storm. When inclement weather is on the way, the heron predicts it by its swift flight. If it flies high and without haste, it foretells fine weather.

The ancestor of this house gave admirable encouragement to honorable deeds. In this his virtuous descendants, as everybody knows, still follow him today. They esteem virtue so highly that they always act cautiously. They despise flattery and all the empty talk of idle people. They look for truth, which they love.

All members of this family have natures which cause them to act cautiously. Above all, they inquire into a subject and, avoiding harm, ponder future cases and possibilities.

They detest all pride and haughtiness and never loathe the poor and miserable. Their jewels are true Christian teaching and the love of the words of Christ. They are contented with what God has deigned to grant them and keep themselves to themselves, doing no harm to other people. But as to those who harm the poor, they always oppose them out of their innate virtue.

If a rapacious adversary wants to harm them, each of them knows how to offer the foe a mighty resistance. They know by nature, when necessary, how to scare away and vanquish heroes of haughty thought.

To the Most Honorable Andrzej Kochcicki, Baron of Kochcice, Lubliniec, Koszęcin, and Turnawa, etc., My Very Gracious Patron A4

Calliope of Parnassus, my fair Muse, who used to make learned rhymers drink the sweet water of the Hippocrene fountain, gave to me also a pen with which to write these simple rhymes.

With this I have written of the first workshops in Lebanon, which Tubal-cain son of Lamech built in the land of Syria before the Flood, and in which he forged copper and iron.

Then of who began this work after the Flood, in the world's second age, who was the first forgemaster, and in which country the forging of iron arose.

Also about how the work of a forge should be ordered, and how to manage the practice of such work, how to deal with forgemasters, and how to erect the forges.

Next, the appearance of various ores is described, and the countries in which the iron is best, without omitting an account of how this noble craft of ironwork A4v
was brought into our regions.

Next comes the power and strength of good iron, the properties arising from its nature, how its virtues surpass those of jewels, how it brings glory to good people and harm to evil ones.

At the end is a picture of the forgemaster, of his practice and customs, his heavy toil, his pains with the fire, his poor and miserable life.

Having gathered all this in my book as you see it, I dedicate it to you, my virtuous Lord. Let the book be spread under the protection of your noble arms among forgemasters everywhere.

Pray accept this little book from me with gratitude. Condescend to endure its simplicity, since it is to you I have dedicated this writing of so little value, which might rightly be despised.

I have done so as a sign of my goodwill toward you, as a remembrance and as a token that I am glad to serve you and that I wish you every good.

Forgive me for what is unseemly, as you should do since you are compassionate and I am untrained, not on account of this poor gift but in consideration of the courteous endeavor of your well-wishing servant.

You deserve great thanks from me, my worthy Lord,
B for your goodwill and for the benevolent care you
have taken of me. This has obliged me to serve you.

Your kindness causes people to pay homage to you, because you display it to everyone. You do not discriminate between people but know how to respect poor and rich equally.

All this comes from your virtuous nature, which brings forth such noble actions and is accompanied by reason, great heedfulness and learning, and above all by fear of God, the most important attribute of all.

In our time of confusion of creed it is a fine thing to have a steady heart. You are not ashamed to confess Christ before the world, Who is your prop and your staff.

You act as an upright Christian, despising worldly devices. On this you and all your house rest, constantly augmenting God's glory.

Live, then, to the great adornment of your country, so that foreign people, too, may take you as a pattern in honorable entertainments! May constant good fortune be yours for many years!

My benevolent Lord's
most humble and
obedient servant
Walenty Roździeński.

Bv

THVBALKAIN filius Lamech primus inuentor æris & ferri.

Laus hæc prima tibi Thubalcaine
Debetur, rigidi repertor æris,
Dum primi facies nitebat orbis
Qui post diluuio periit nefando.

To you, Tubal-cain, the highest praise is due as the inventor of rigid bronze at the time when the first world shiningly appeared, later to perish in the Flood in an unspeakable manner.

B2

On Who Were the First Smiths in the World to Discover Iron and Copper and the Different Crafts by Which They Are Made Serviceable through Treatment with Fire

Gold and silver were not yet known in the world when there came iron and copper, which God gave to the first man for his needs after his Fall,

so that he might manage his husbandry, dig and plough his soil, and do all that was needful to procure his clothing, food, and bread.

Josephus, *De Antiquitate Judaeorum,* vol. 1.

It was Tubal-cain who first taught his children to forge with a hammer, to use blast in the fire, and to smelt. This he did before the great Flood had drowned the earth.

The high mountains of Hermon, which are part of the mountains of Lebanon, going from there to the south and separating the land of Basa from Coelesyria, encompass Galilee on the north and the south.

To this purpose he first procured for himself new implements and driving mechanisms, which he set up in Syria below Mount Lebanon on a beautiful site overshadowed by the rocks of Gilead and of Hermon.

Cain the fratricide, son of Adam (Gen. 4). Cain means possession.

Thus it happened that all the sons of Adam, after his expulsion from Paradise, lived at the foot of the mountains of Lebanon where Cain built a town and dwelt with his offspring.

B2v

Enochia, the first town, was founded by Cain before the Flood.

As the erection of the town began at the time of the birth of Cain's son Enoch, Cain ordered it to be called Enochia.

According to Berosus, Naamah was the sister of Tubal-cain.

There the clever woman Naamah dwelt, she who was first in the world to make silk attire, as in those days the luxury among Cain's sons was great.

Damascus means drink Sakah's blood or, perhaps more likely, drinking from a sack filled with blood.

It was at Enochia, or nearby, that the sons of Noah built soon after the Flood a town which they called in their language Damascus.

This name refers to the blood of Abel, who, as is told, had been killed by Cain at this place; it may be explained as a lump of soil drenched with human blood, or rather a sack filled with blood.

The name Enochia, derived from Enoch, may be interpreted as a town dedicated and consecrated to

God, though despite its fine name it was full of cruel and very wicked people.

This explains the custom of today's evil people who adopt fine titles while slandering the good people they despise by calling them by disgusting names.

According to the evidence of Moses, this forgemaster was born of Zillah and begotten by Lamech, of the seventh generation after Adam, of the lineage of Cain who killed his brother Abel.

Lamech, a descendant of Cain, was a tyrant and the inventor of polygamy (Gen. 4).

The people of this nation were cruel; and at first giants only were born among them. Fond of ferocity, they were the first who fought with the sword.

B3

Josephus, *De Antiquitate*, vol. 1.

They were also the first to invent all knightly implements for fighting and shedding blood, some of which were hammered of steel and some of hardened copper.

Confident in their arms, they forsook God and attacked peaceful people, the descendants of Seth, whom they robbed of their goods and in this way entered into possession of land by the sword.

This might have been predicted from their father's name, Tubal-cain, which means possessor or sovereign on earth, a man huge and commanding.

Tubal-cain is a name of presage. He was a brother of Tubal, an ironsmith in the territory of Asher (Gen. 4).

Moses calls him Choresch, which shows him to have been a smith—an artisan in a craft using fire—because the Jews called those who forged with fire *choresches.*

Although those forges had been swept away by the Flood, nevertheless Noah's descendants kept them in mind. They were discovered by Tubal, Japheth's son, a very clever smith, who found ore on this spot,

seeing it exposed after the waters of the Flood had removed the overlying earth. There he rebuilt the forges, imitating in this art the first master.

Others say that Japheth, seeing that his son Tubal was skilled in metal crafts and mining, showed him those places below Lebanon where Tubal-cain had built his workshops.

B3v

It happens also in our regions nowadays that such ores are found in marshes, just beneath the surface, where the water removed the topsoil in those days when the world was flooded.

In some places there occurs a large amount of such subsurface ore, only it must be taken out of the water. Wherever it is found, it lies in bogs, and it therefore is popularly called bog ore.

Amidst these forges lay Mount Lebanon, overgrown with cedars and incense trees. The mountain was surrounded with land of high renown for its abundance in oil, wine, and corn.

Cabul means a sandy, barren, and unattractive land (Joshua 19).

However, the forges were located in barren sandy or marshy places. For this reason Hiram, King of Tyre, named the land Cabul, which name is still in use.

These lands were given to the descendants of Asher after their exodus from the bondage of Egypt, their inheritance as distributed by their leader Joshua, by lot and by the word of Moses,

Deuteronomy 9.

who, feeling the approach of death, divided in advance the possession of the land, and told them of its wealth of iron, copper, and all things.

I Kings 9.

B4 This province had twenty towns, and there were still many forges here in Solomon's times; Solomon granted the province to Hiram, King of Tyre.

As the price of the iron hardly exceeded the cost, Hiram changed his mind and returned the province to Solomon, saying that the land of Cabul was barren.

Joshua 19.

Among the towns of this land there was the old town of Cabul, exceeding all others. It was surrounded by workshops and ore mines.

It is told that king Solomon and Hiram stayed in this town, when they were surveying all the land of Cabul that had been granted to Hiram.

This land lies within the bounds of pagan Galilee, which is near Phoenicia and has a common frontier with Sidon and Tyre as far as the sea, where Zarephath lies to the north.

Zarephath lies on the seashore, like a vanguard against Sidon, and was once inhabitated by merchants of Asher's offspring dealing in copper, iron, and glass.

For these reasons the Jews commonly called the town in their language Zarphatah, which means smelting house and glass furnace, where glass or metal are molten.

It is said that nowadays this town has but eight houses not destroyed, the town is demolished to such a degree; yet the ruins show how extensive it was and how vast its marketplace.

B4v

Henricus Rutin, *Liber Itinerarium Sacrae Scripturae.*

There still stands the undamaged cell where Elijah, the holy prophet, raised from the dead a widow's son, as is told by the people who dwell in that place.

Beyond the town gate, on the southern side, a chapel is shown on the spot where the prophet called the widow, who was gathering wood, and asked her to give him water to drink and bread to eat.

These Cabulian forges were still standing in the days of the Apostles in the valleys of Lebanon. Besides iron, copper was forged and excellent brass still produced.

Apocalypse 1.

The Jews had in their country no other metals besides copper and iron. Yet they got pure gold and silver in unlimited amounts from the Arabs and from the land of Ophir, where there were gold mines.

Ophir, which means the land of gold, is in India.

C

VVLCANVS siue Mulciber informis, filius Iouis Cretensis, ex Iunone Deorum Faber.

Quem peperit Iuno Saturnia semine magni
Fœta Iouis, ferri ostendit mortalibus vsum
Mulciber informis, dum cœlo extruditur alto,
Excipit & lapsum, contuso poplite, Lemnos.

He whom Saturnian Juno bore, pregnant with the seed of mighty Jupiter, and who revealed the use of iron to mortals: Mulciber the deformed one, who was hurled down from high heaven and taken in by [the isle of] Lemnos as he fell and shattered his knee.

On Vulcan, the Forgemaster of Lemnos

Cv

After the deluge, in the second age of the world, Mulciber or Vulcan, who was a man of great wit, restored the forges that had been inundated and destroyed by the Flood.

Cicero, *De Natura Deorum*, book 3.

It was he who first invented the craft of ironwork, of which nothing had yet been heard at that time. He was begotten, it is said, by Jupiter on Juno, Saturn's daughter.

Hesiod in the *Theogonia:* Juno, united in love, gave birth to the famous Vulcan.

But Vulcan, ill-fated in his youth, soon experienced his mother's awful cruelty; as soon as he was born, she flung him down on the isle of Lemnos, in order that he should perish there.

His mother cast him forth so pitifully because he was very wretched-looking after his birth. When he fell from aloft, poor Vulcan's legs were so badly bruised that he was crippled.

It was Thetis who helped him after this fall to earth; she took him in and ordered her sisters to foster him. When he grew older, he became quite famous and popular throughout the world.

Homer, in the hymn that he wrote to Apollo: Vulcan himself is my lame son, I myself gave birth to him and, having grabbed him with my hands, flung him into space. The venerable Thetis, daughter of Nereus, then took him and went to her sisters to whom she gave him to be brought up. C2

Others write that when his mother Juno threw him down to Lemnos on account of his awkward appearance, a she-ape living at the place where Vulcan fell to earth reared him with her young,

or that it was by Jupiter and not by Juno that he was so miserably cast down on Lemnos, so that afterwards, when he had grown to manhood, he always tried to revenge himself on Jupiter.

Homer in the *Iliad,* Book I: I was thrown by the foot from heaven by Jupiter and barely saved myself, for after flying over heaven along with the setting sun and falling on Lemnos, I had scarcely any life left in me.

It is supposed that there were four Vulcans, yet all that the others did has been ascribed to the one from Lemnos; he alone has gained fame according to the general opinion.

Cicero, *De Natura Deorum,* book 3.

The first Vulcan was the one born of Minerva, raped by Apollo. The second was called by the Egyptians the Fat; he was appointed guardian of the land of Egypt.

C2v **The third was the one born of Juno, who was cast down on Lemnos and became the first smith. The fourth held those isles near Sicily, called volcanic.**

Plutarch, *Pelopida.*

This latter Vulcan begot many sons on various goddesses and mortal women, but only one daughter, Hermione, whom Venus bore to him.

Homer; Virgil, *Aeneid,* book 8; Macrobius, *Saturnalia;* Piso; Isatius.

Some say that his wife was that Venus who was begotten by Jupiter of Crete on Dione, others say it was Majesta, and still others that it was Maia, but some say it was Aglaia, one of the Graces.

Vulcanus, facifer in nuptiis.

This Vulcan first initiated the custom of conducting brides with burning torches to their marriage beds. He was therefore called the god of fire and had Lemnos consecrated to this.

Euripides in *The Troades:* Thou bearest torches at the weddings of mortals, O Vulcan!

This manner of lighting, invented by Vulcan, still continues in our country among people of honorable rank, who use candlelight when leading young couples to their bridal beds.

Vulcani Lampadophoria.

Great honor was given to Vulcan in those bygone days, and a tournament of fire was devised for the purpose. Taking torches in their hands, the people raced to a goal, doing this on a solemn day dedicated to Vulcan.

If during the race someone's light went out, he had to give up his place at once and was railed at; but glory was the lot of the man who first reached the goal without his light going out. C3

If someone from the rear succeeded in overtaking the one running at the head and reached the goal first, then the vanquished runner, shamefaced, had to pass his burning torch to the victor.

Lucretius in Book II: And like runners they hand on the torches of life.

Some say that this race was run to commemorate Vulcan's discovery of fire.

Hence they consider that every craft practiced with the use of fire was taught by Vulcan and that he created many fine things, both in ironcraft and in the goldsmith's trade.

In the goldsmith's art his first masterpiece was the crown which Venus, and later Ariadne, wore on her head; there were also bracelets and many other fine pieces of workmanship.

As to ironwork, he forged various pieces of armament for Jupiter when he fought the giants, defeating this tribe of Titans and then hurling it down from Olympus to Earth.

Hesiod in the *Theogonia.*

Tzetzes in the *Historia,* verse 335 of the tenth chiliad: Vulcan, a certain Egyptian in the time of Noah, who is also called Noah, Dionysus, and Osiris, discovered fire and invented all the arts arising therefrom.

On Lemnos Vulcan first began to forge with a hammer, and some say that he did this on the hillock where he had been thrown, on account of his ugliness, by his mother Juno. C3v

This hillock, somewhat higher on the side toward the east, is on the seacoast. Here was built the town Hephaestia, so called from Vulcan's name.

The Greeks commonly call Vulcan in their own language Hephaestus, from the word for fire, and still use the name Hephaestia for this spot on Lemnos.

This town is mentioned by Ptolemy in book 3 of his *Geographia* and by Galen in book 9 of the *Simplicium Medicamentorum Facultatibus.*

This hillock still stands all barren, quite dried up by Vulcan's fire. It is said that no tree has grown on this arid spot from that time to the present.

This is the *terra sigillata* of Lemnos, described by Galen and Dioscorides.

However, a very precious earth is dug in this place, which is considered an excellent antidote to any poison; it has been described in detail by some eminent medical men.

It is also said that on this very island fire was first discovered after the Flood; this was attributed to Prometheus, to whom alone the glory was ascribed.

But those who worked with fire believed that all such
C4 crafts come from Vulcan, whom they considered the
first master of all crafts using fire.

The emender [?] of Sophocles, from a statement by Lysimachides.

The people therefore erected a monument common to the two and portrayed both their faces on it. The figure held in his hand a royal scepter, like a king, and people paid him divine honors.

Hellanicus in his commentaries on the building of Chios.

It is also told that the first iron armor was forged on this very island. This craftsmanship was ascribed to Vulcan alone, and the first praise for this was given to him.

Still others think that Vulcan's forges were in Sicilian Etna, where the Cyclopes forged steel cuirasses and arrows that flew blazing with a terrible fire.

Hesiod in the *Theogonia;* Virgil, book 7.

Jupiter was clad in this armor when fighting the giants, and he defeated them with these arrows; one of the arrows mortally wounded Aesculapius, who was accused by Jupiter.

When forging, Vulcan always kept at each of his workshops these Cyclopes—men of tall stature, fierce and monstrous, but skillful masters in the craft of forging iron.

Their skill was first proven by making two hard-steel cuirasses; these they made at Vulcan's command according to the wishes of Thetis and Venus.

The strength of one of the cuirasses was tried by Achilles near Troy, when he fought for beautiful Helen. The second was used by Aeneas when fighting Turnus for Lavinia, daughter of Latinus.

Phylarcus in the *Rebus Fabulosis.*

Hermione was given by Vulcan a gold necklace made in this same cyclopean workshop, but it was unlucky for it led to her doom; Vulcan would not acknowledge his daughter.

C4v

Hesiod in the *Theogonia.*

Later, because of his skill, Vulcan came to be regarded as the god of fire; a feast day was celebrated in token of his fiery potency, and a lion was burned in his honor.

The first forge for the craft of ironwork was built by Vulcan on Lemnos, and the iron-making crafts were next brought to some of the Sicilian islands.

Agathocles, *De Arte Ferraria Vulcani.*

Two of the forges there were larger—those of Hiera and of Strongyla—but ironworking has now stopped in these places. For this reason all those islands were called in Vulcan's honor the Vulcaniae.

It is told that some of the rocks on these islands used to give out fire like Etna, so intense that water dried up in whatever place was reached by it.

Apollonius of Rhodes in the *Argonautica,* Book 4:
He arrives again in lands where anvils ring as they are struck by Vulcan's massive hammers and where waves of smoke rise with heat.

Of old, when Vulcan's isle Lipara was called Meligunis, it was no less wealthy than Sicily. Rich in alum, it had plentiful and valuable hot springs in addition to wealthy forges,

of which there were plenty on the island, very famous for the belief that the Cyclopes had forged there in olden times; Vulcan had mastered this island and kept Cyclopes there.

D

It is said that some of the rocks burned perpetually with a sulphurous fire, crashing with a great noise like ghastly thunders, much like the Silician Etna.

The crashing coming through holes from underground

made it seem from afar as if noisy forgemasters' bellows were violently blowing their furnaces, and as if turning wheels were hammering the metal.

Just as on Lemnos, this gave rise to the belief on this island too that Vulcan had been sent by Jupiter from heaven as god of fire for the craft of ironwork on earth.

Callimachus in the *Lavacrum Dianae:* She then sought out the Cyclopes: she found them on Lipara, where they were dwelling. The island was then called Meligunis. They were standing around white-hot blocks of iron lying on Vulcan's great anvil: their huge task urged them on.

There is an island near Lipara—extending on one side almost to the shore of Sicily—that is famous for its subterranean caverns from which inextinguishable sulphurous fire once furiously burst.

Navigation was very dangerous when that island was raging with the fire, which very often burst forth from the densely serried rocks in desert places.

Dv When the wind rushed on the fire, it drove the flame and smoke with a terrible uproar toward the sea. The immense heat caused the water near the shore to dry up wherever the flame reached.

However, the fire would so rage only by night; by day it would just smoke or fume so much that you could not see Sicily's shore even from nearby.

For this reason and according to heathen use the island was consecrated to Vulcan; here also he had his forges, where dwelt those same Cyclopes who were forging at Lipara.

Men of huge stature, naked giants, foremost in the craft of ironwork, served Vulcan. These Virgil mentions in his poems, especially Steropes, Brontes, and Pyragmon.

Each of the Cyclopes got his name from his own special craft: the first, Brontes ["thunder"], got his name from the crashing of hammers and the roar of the bellow's blast;

The second, Steropes ["lightning"], was named for the frequent flashes at night from the blazing fire; the third, Pyragmon ["fire-anvil"], got his name from the heated anvil and the great fire. Albericus.

Virgil, *Aeneid,* Book 8: Near the Sicilian shore and the Aeolian Lipara there rises a steep island marked by smoking rocks. Beneath it is a den with caves of Etna, originally excavated by the furnaces of the Cyclopes, reverberating with sound. Powerful blows on the anvils are heard, quenched steel hisses in the caves, fire crackles on the hearths. This is Vulcan's home, and the land is called Vulcania. It was here that the Lord of Fire descended from high heaven. D2
In a vast cavern the Cyclopes were forging iron—Brontes, Steropes, and naked-limbed Pyragmon. A thunderbolt such as the Father often hurls down to the earth had been shaped and was partly polished.

Though the Cyclopes were the offspring of a royal family, yet they practiced the craft of ironwork. Theodontius says they were born of Amphitrite by Neptune.

In his turn Neptune was born of Opis or Rhea by Saturn, of the same wife who bore Cretan Jupiter. Thus they were closely related to Vulcan.

However, they were huge and terrifying because they were born with a strange form, differing from men: each of them had but one eye in midforehead, which made their faces look immense and terrible.

Although born with such a strange form, they had great wit and clever minds and were invincible in strength and power, which made them resemble gods.

Apollodorus, Book 1: After these the Earth [Gaea] bore by Heaven [Uranus] the Cyclopes Arges, Steropes, and Brontes, each of them having only one eye in his forehead.

Hesiod in the *Theogonia:* She next bore magnificent Cyclopes, the splendid Arges, as well as Brontes and Steropes, who made thunderbolts and thunders, the mighty arms of Jove. They resembled gods in every respect, except for having a round eye in the middle of their foreheads: hence their name.

D2v

CYCLOPES VNOCVLI FRATRES

Filii Neptuni ex Amphitrite egregii mineratores & artifices in arte Vulcani.

Gnauiter accensis operas fornacibus vrgent
Et conflant igni mortalibus vtile ferrum
Vnoculi fratres, monstrosa propago Cyclopum
Brontesque Steropesque & nudus membra Pyracmon.

Zealously, their forges kindled, they press on with their work and with fire smelt the iron that is useful to mortals. They are the monstrous race of Cyclopes, the one-eyed brothers Brontes and Steropes and naked-limbed Pyragmon.

On Certain Smiths to Whom the Discovery of Iron in the Second Age of the World Is Ascribed

D3

Opinions have always differed as to who invented the noble craft of ironwork in the second age of the world, or who taught men to forge iron and to build forges.

Diodorus says that this art was invented and spread by the Idaedactyli. This might be true because with their skill in the craft of ironwork they surpassed many a forgemaster.

Diodorus, Book 6.

These smiths lived in Phrygia beneath far-famed Mount Ida at a place called Bereemthus. They were commonly called Idaedactyli after this mountain.

Hellanicus.

From the songs of that poet who wrote the *Phoronide*: There in their mountain homes dwelt Celmis, Damnameneus, and powerful Acmon, Phrygians from Ida, skilled tricksters, excellent metalworkers, and servants of Adrastea. They were first to discover the art of the wise god Vulcan in the high mountains, and they taught how iron put into fire may be given various forms.

Others have ascribed this to the Cyclopes, who must have first come to forging iron. Still others consider the Chalybes to have been the first inventors of this art, before the Cyclopes had mastered it.

Plinius, Book 7, Chapter 50.
D3v

Others again believe that forgemasters of a greater skill had of old their workshops on the famous isle of Rhodes and that they were called the Telchines.

Strabo, *Geographia*, Book 14.

But it seems more probable that this honorable art was invented directly after the Flood by the descendants of Noah, particularly by Tubal, a smith of high rank in this noble craft of ironwork.

He restored the forges, destroyed by the Flood, which had been built by Tubal-cain, Lamech's son, who first found the mines whence ore was taken.

Apocalypse 1.

This Tubal, having come to Europe to forge after this part of the world fell to the share of Japheth's

progeny according to the division among the brothers, started by constructing ironworks and mines.

Genesis 10.

He was named Tubal by Japheth, who predicted that in the future he would be as famous a craftsman and master in ironwork as Tubal-cain.

This came to pass, and since Tubal means an ore miner and the name smith comes from the blowing and forging of iron, it is all one, for a smith can never produce iron without a miner.

The learned think that Tubal invented the smelting of metal, since among his descendants there were always people practicing metal crafts.

D4

Pomponius Mela, *[De Chorographia]*; Josephus, *De Antiquitate Judaeorum*, Book 3, Chapter 2; Berosus, Books 4 and 5, *Antiq.*

A town in Baetica was named Tubal because it had been built by Tubal, from whom the Spaniards originate—as Josephus and Berosus tell.

Hence the said Josephus names Spain Tubal, as do the Jews, because Tubal was the first to build forges there and to forge steel and iron.

Barzolon means "iron-town."

Although this happened long ago, traces still remain in Spain where Tubal's workshops stood. On the seacoast there is a town called in Hebrew Barzolon.

It got this curious name from the iron which was forged there, which had formerly had so great a renown that there was said to be none better in the world, except in Colchis.

This was because the smiths forging there were Chaldeans and Jews, or else Carthaginians from Sidon; others say they were old Chalybes.

In any case, they all talked Chaldean, though differing in many words. They were all descendants of Tubal—both the Carthaginians and the Chalybes.

Iberia is a region between Colchis and Armenia.

Before settling in Spain, they smelted iron in Iberia and in Colchis, those islands in the Euxine Sea, near the shores of the Chaldean kingdom.

This means that they came from Chaldea, whose name
D4v derives from the word for steel, for *chalybs* means a

piece of hardened iron; the knowledge of how to harden iron was their own.

Homer calls them Alizoni; later Strabo says they were Chaldeans, since they lived in Pontus near the river Thermodon where they also forged steel.

Strabo, Book 12.

In Spain there is a river still called Chalybs, from those Chalybes, because Justinus locates these smiths there. This place still produces excellent iron.

Justinus, last book.

Certain tests, they say, show that the iron of no other country is sharper. It is obviously the water in which the smiths quench the iron that brings this about.

This power is ascribed to the water of the rivers Bilbilis and Chalybs near which the old Chalybean forges still stand today, where iron of so noble a quality is forged.

Its strength and superior quality may be seen in those needles which in our country are called Spanish. It will pierce any other kind of iron without harming itself or losing its nature.

Justinus in the last book: The iron here has an especial substance, but the water is stronger than iron itself; quenching in it increases the sharpness of the iron; no weapon will meet with approval unless it has been dipped in the river Bilbilis or Chalybs. Hence the people settled on this river are called the Chalybes and are considered to surpass all others in the quality of their iron.

Virgil says they went naked when making iron in their workshops. I do not wonder, as few do not do so, and this is still the custom among smiths.

E

Herodotus writes that those Chalybes were circumcised like Jews, observing this custom since the time of Abraham, whom they followed as a father.

Herodotus, Book 2.

When the noble craft of Vulcan became greatly famous the world over, others began to imitate the master, each endeavoring to outstrip the others.

Some did their digging deep into the earth for fame alone, wanting their names to remain known forever;

others did it from avarice, wanting to become rich and acquire wealth.

They hammered hard rocks, broke stones, and took no care of their health, looking for ores in the ground. It was an unusual human endeavor to look for such things at that time.

Ovid in the first book of the *Metamorphoses:* But they have delved down into the bowels of the earth and, approaching the Stygian darkness, have started extracting the riches there concealed, incentives to crime. Pernicious iron and still more pernicious gold have already appeared on earth.

Plinius.

Ev **Cadmus, son of Agenor, when sent by his father from Phoenicia to Greece to look for Europa, who was wrested from him by Jupiter, taught the Greeks the smelter's art,**

which he himself had learned in his youth in the land named Cabul near Mount Lebanon from his neighbors the Assyrians, famous Jewish masters in this craft.

I Kings 18.

For their forges stood near Phoenicia, not far from Sidon and Tyre, near rocky Mount Carmel, where in olden times Elijah defeated the false idolatrous prophets of Baal.

This Cadmus, having left Phoenicia, appeared in Boeotia and went unnoticed at first for he had not a king's but a miner's manner, having divested himself of his royal attire.

There he found ***lapis aerosus,*** **the cupreous stone now called cadmia, near the town of Thebes, and he displayed an unheard-of skill in smelting it into copper and then into brass.**

Boccaccio in *De Genealogia Deorum,* Book 2, Chapter 63: Pliny, in his *Naturalis Historia,* says that Cadmus discovered a stone quarry near Thebes and invented the arts of working gold and of casting metal; Theophrastus maintains that he did this among the Phoenicians, but long after the time indicated, for what was written about him first occurred about the year 3805 from the beginning of the world, whereas the latter was about the year 3838.

Vulcan's tribe, having left those isles and the Greeks, came first to Slovak lands, to Serbia and Bosnia and to Hungary, where the people at that time were everywhere Slovak. E2

In a very short time they had set up many an ingenious forge, but when these places passed under the Turkish yoke they were abandoned one by one.

The Slovaks later brought the ironsmith's craft from Hungary to Carinthia and into the Styrian lands and taught the Styrians iron- and steelmaking.

The ironsmith's craft was later brought over from Styria to Italy by Germans; the Italians named the town where iron was first forged Ferrara, a name derived from their word for iron.

The goddess Ceres brought the craft of ironwork into German lands during the reign of King Swab in Germania, after Hercules had died in Spain, when she was looking for her daughter Proserpina.

From the *Librum Exemplorum,* first precept, page 36.

The Swabians therefore paid her great honors and called her the Lady of Iron. Others thought her to be one of the goddesses and addressed her in their adventures.

In remembrance of her reaching them by sea in a boat, and for the sake of the benefits they received from her, she was represented standing in a boat and the people brought her sacrifices.

Later she left the Germans and went to Italy to teach husbandry to the people there. For this she was named there the goddess Ceres and called queen and mother of all goddesses.

Aventinus, *Chronica,* Book 1.

Tacitus states that in his time there was no metal in Germany besides iron, which was then forged only in Hesse, in a single workshop, E2v

which a Roman commander-in-chief, Curtius Rufus, acquired when he was sent with troops by the Romans to repress the insolent Germans, who were doing much harm to the Romans.

It is said that the old Goths, who are also considered

true Germans, had had forges since very early times. The iron they forged was so good that it equalled Noric iron in quality.

There still exist forges in Sweden in which good iron is smelted, to which the ironsmith's craft was brought by the Chalybes from as far as the island of Colchis.

Plinius, *Naturalis Historia*, Book 34.

The first forges in Germany were those at Noricum and at Passau. They were mentioned by poets even before the coming of Christ and praised on account of the good grade of iron smelted,

and also on account of the reliable and hard weapons, swords, and armor that were made of that iron. It was said at the time that there was no better iron in Europe.

Botak, of royal lineage, the first Bohemian forgemaster, started forging iron in Bohemia in 677 A.D.

During the reign of Krok in Bohemia a certain Botak, an elderly man, set out with his kinsmen and servants to look for fertile lands in the wilderness.

One day, having come to high mountains, he began to dig the earth with a mattock, looking for fruitful soil.
E3 He suddenly struck very good iron ore, of which there was plenty everywhere.

Haiek in his *Chronica*.

Having taken samples of the ore, he joyfully returned home and gave it as a present to Krok, his master, who, taking the ore, immediately ordered him to go to those mountains and to set up workshops there.

Krok appointed him prime overseer because he was the first to discover this ore and so should be the first master of the works.

When Botak reached these mountains with his artisans, he first built a house for himself below the mountains and then constructed forges and named this place Zdechovice.

The place is called the same today. A village is there, and the house is still standing. The mountains where the ore was first found were named the Iron Mountains.

There have been workshops in Bohemia ever since

Lubliniec and is now within the territory of Lord Andrzej Kochcicki.

These three neighboring forges along the river—Bruskowska, Plaplińska, and Pusta—are in the district of Lubliniec in the Koszęcin territory in a place favorable for the ironsmith's craft.

Here the marketing of iron is easier and the sales better than in all the other Silesian forges. But, as I have already written, the antiquity of the Bruskowska forge exceeds that of all the others on that river.

How old it is may be ascertained from the slag heap, which is so high that the smelting house cannot be seen on one side, but only the sparks flying out of the chimney.

Fv This forge was also abandoned and stood for a long time unused because it was not possible to secure a skilled forgemaster.

Then Biernat, Duke of Opole, called to this forge one Glawer from the land of Meissen, a famous forgemaster, skilled in constructing the driving mechanisms for ironwork.

He at once repaired the collapsed machinery and set up the forge in its original state. To this day iron is forged there, smelted from very good Bytom ores.

The Poles first learned how to work in iron from these Germans and in this way took over these crafts from them, so that in their speech they still use the [German] name for every tool.

Long ago there were in Silesia no power-driven forges such as there are now on every river. Forgemasters acted as hand smiths, blowing the bellows themselves and hammering by hand.

trabo places Moscovy djacent to Colchis.

They worked as the ancient Chalybes used to work in Colchis, whom the Moscovite frogemasters called Sfistones also imitated, smelting iron as the Chalybes did.

After Glawer had left the world, the Freszels soon

those ores were discovered, but their equipment for ironmaking differs from ours.

In our country the ore is smelted in low, round furnaces which produce iron rapidly, while there the furnaces are great and spacious, high like chimneys, and built square.

It is therefore hard to introduce a tuyere with a large hole through which the bellows can blow from above into the hearth. This causes much charcoal and iron to burn, so that it flows like mush through the slag notch.

Next, there is one workman who is constantly busy breaking the slag in a stamp mill, and hence performs needless toil, picking the burned iron out of the slag, which is again thrown into the furnace after being mixed with [new] ore. E3v

The blowing in those furnaces is slow, for the ore needs to be in blast for as long as three days to make a bloom, which has to be pulled out of the furnace with a chain, as out of a well, and smelted a second time.

Long after the Bohemian forges came the forges of Meissen; they were erected for the first time by Otto, Prince of Saxony, sovereign of this Saxon country.

These forges arose during the reign of Otto the Second, Prince of Saxony, about the year 974 A.D.

The iron and copper crafts were brought to this part of Silesia from the province of Meissen by some wandering German forgemasters, skilled artisans in both these crafts.

Having arrived here, they at once scattered with their implements all over the Silesian land. Some of them set off for the Sudeten Mountains and there looked for places proper for their crafts.

Having found plenty of ore below the Riesenberk Mountains, they immediately erected forges nearby, on the plains near the old town of Szmidberk, so named on account of the forges.

[The Ryphaean Mountains may be the Karkonosze or the Riesengebirge or perhaps the Carpathians.]

Coppersmiths built their forges among the Ryphaean Mountains, so called because the Ryphaean

Sarmatians came into Bohemia this way with the Czech language.

These forges were located near the old borough of
E4 Kupferberg on the bank of the river Bober; they were properly constructed in a very brief time, to the number of seventy.

Now they are all abandoned. The only marks remaining on this place are cinders and hammer scale; hence the place where this copper was forged is called Szlakengrunt on account of the slag found there.

Of these seventy forges on the river Bober, one workshop still remains, where, it is said, copper is at present produced which is as good as that made in Hungary.

Seeing that the smiths of Szmidberk had good luck, other forgemasters were attracted to the region, some of them settling in the duchy of Legnica, while others remained in the duchy of Żegań and still others set out for the duchy of Opole.

In this way, many forges were soon set up in Silesia. Some of them are in the south near the Sudeten Mountains, others in the east or north near the Kingdom of Poland.

The duchy of Legnica was very famous for its forges in the year 1242 when Henry, called the Pious for his virtues, was killed by the Tartars.

While fighting the Tartars he was assisted by a large retinue of chosen metal craftsmen, all of proven valor, who defended him bravely, though they all perished.

It was not iron or copper alone that was worked in this duchy; they also had gold, which they gathered from the sand of the river Kath near Goldberg and melted in a forge.

E4v Again, the duchy of Żegań was not lacking in good ironwork shops, especially at the time when Duke Henry was ruling there, he who was called Ferreus.

In 1337 A.D., Henry, Duke of Żegań, surnamed Ferreus, constructed forges in his dukedom.

It is said that he was so named by the common people because he set up in his duchy certain ironwork held the forge, Germans also, two full brothers, who were sent to the forge by the duke.

One of them, being, as people tell, an unjust man and a usurer, covetous of other people's things, was punished for his impiety by God; he was sent to Satan, F2
who stripped his skin from his body.

He died the ninth day after the cruel flaying, and his brother immediately fled from the forge, being in debt. Having paid off the debts, the duke transmitted the works to Hercyg as recompense for his services.

The highway leading from this forge to Lubliniec was called Hercykowska, after this very man. In witness of this is the grant of the Dromowskie forests, owned by the citizens of Lubliniec.

The Roździeńskis and the Bruseks are descendants of this family, which on its first arrival in this country was called Hercyg. Later, when a change came, one of them was called Brusek, from Brutus.

From that time until now they have called themselves the Bruseks, although it would be more reasonable to call themselves Hercyg. They still hold in this forge the native place of their ancestor Hercyg.

It is an old family and an ancient house. From its brave heart it is called Serdeczny [i.e., hearty, cordial], which expresses their great intelligence and vast attainments and denotes all their good qualities, as you see.

The arms show a heart within a band in a white field; above is a helmet from which, on the two sides, there issue the outspread wings of an eagle, on each of which is a heart.

F2v On the Aforesaid Arms
Called Serdecznik

Although Fortune is deceitful, an audacious heart, scorning her, is tempted to do great deeds. It carries off the victory, the result of upright heroic virtue and the work of a righteous heart. To trust in God in an adventure brings a hearty result; three hearts mean the everlasting triumph of virtue.

F3 The forgemasters from the land of Meissen, seeing in our country forests in plenty, streams with unfailing water, and an abundance of ore, constructed forges in every suitable place.

After the Małpadewskie forges came the ones at Rybna, but some of these have already been abandoned and now only the notable Tworogowska and Kotowska forges remain still working on the river Rybna.

These forges are located in the territory of Swinów in the lands of the lord of Reder, in the domain called Toszek; at present they outmatch all the forges in the duchy of Opole in the abundance of their forests.

Later three more forges were erected beyond these; two of them stand near one another on the river Kłodnica—one in the domain of Pszczyna, the other in that of Bytom, lower, in the territory of Kochłowice.

The third forge is also in the domain of Pszczyna. It is constructed on the river Roździanka in the estate of Bogucice, and now it excels other forges in its rights, privileges, and forests.

However, owing to some sorcery, it was once abandoned and so stood idle for several years, until the forgemaster Jurga, called Kleparski, repaired all the damaged machinery.

He was brought from Klajneberk by Kazimierz, Duke of Cieszyn, who settled him there on that place; this is proved by the right to this forge granted to him by the duke.

Finally, the Roździeńska forge was constructed; it was by no means the worst of those three in the domain of Pszczyna. It was set up in the Roździeń territory on the river Roździanka, like the Bogucka forge. F3v

Later the craft of working in iron was brought from Silesia to Poland by certain men from the land of Meissen. One of them, called Warchoł ["the squabbler"], erected the first forge near Krzepice on the river Kostrzyna.

People say that it is the oldest forge in Poland and hence call it Stara Kuźnica ["the old forge"]; of this the proof is the great mass of slag, which does not appear in such amounts at other forges.

This forgemaster from the province of Meissen was called by the Polish people Mysznar. He was brought to this region by Przystański, the lord of this district.

There are still six active forges on this river in the Krzepice district, and there are yet more forges in certain places near the forests and on other streams.

After the Krzepice forges came those of Niwka; they were the most famous in Poland on account of the good iron that was made in them; there was no better in all Poland.

Nowadays they still excel other forges here in Poland for their good iron, although some of them were abandoned more than ten years ago because the forests had all been felled.

But as the forests have since then grown large and dense over the clearings, some of the dilapidated or abandoned workshops have been restored.

F4 Six forges were built in a row in the domain of Mrzygłód, and these were called the Niwka forges. Three of them lie on the Czarna Stream near the Pilica highway leading to Siewierz.

The one nearest Siewierz is called Krzemienda; a quarter of a mile further on is the Wargulowska forge. Some say this is the oldest of them all, and some call it the Niwka forge.

The third forge–the Nikiasowska one–lies somewhat upstream; the highway from Pilica runs through the middle of it. On the other side there are three other forges set on the river Warta.

The first one upstream is the Nicowska forge near Zawiercie; it is said to be the oldest of these three forges. Lower is the Marciszowska forge, while the third, a mile downstream, is commonly called Zawadka.

After the Niwka forges came those of Olsztyn, in particular those set on the river Warta, among which there are two excellent workshops, the foremost of their days: Wały and Osiny.

There are still more forges in other places in Poland, some set up long since and some new, many of them in the region of Sandomierz, mostly in the forests near the Góra Świętokrzyska [Holy Cross Mountains].

There is also a copper workshop, erected several years ago in the diocese or domain of the Bishop of Krakow, near the borough called Kielce.

Then this forge is working hard it is said to yield a profit of more than ten thousand złoty from copper every year, in addition to the wages which the artisans take for their work. F4v

Silver ore is also found there, and various kinds of marble quarried; moreover, a delightful azure stone is found which resembles turquoise after it has been dressed and polished.

Other forgemasters, having met in Byelorussia some Sfistones, Moscovite forgemasters, proceeded forthwith to erect in their own country workshops with a water drive, since they had formerly worked their bellows and forged without such machinery.

[Sfistone: a nickname deriving from the Russian word svistoon, *literally a whistler, idler, or lounger.]*

The traces of these Sfistone forges, where they had their implements for hammering and blowing iron, may still be found in Silesia near Opole, near a large pond amidst forests.

Such was the way, as has been written, in which the honorable craft of working in iron and copper was brought to various countries of the wide world by human wit and reason.

For this let God Almighty be praised, who gave to his creatures on earth so precious a metal, by which we gain our food while living in this vale of tears.

G 

Cicero, *De Natura Deorum:* From the caverns of the earth we bring forth iron, which is necessary for the cultivation of the fields.

Do not wonder that I hammer the firm rock so hard by the sweat of my brow: I am induced to do so by the thirst for good wealth, which I expect to gain by digging mines.

The hope thereof governs my digging in the earth, and I diligently watch its exhalations [as signs of miner-
Gv als]. Having pierced through the hard stone with damage to my health, I let myself down to the very bottom of the mine.

Yet my hopes often deceive me, for ore is not found in every place I dig. Many a time I lose the work it has cost me, yet I care not for the cost or the labor.

Then I dig elsewhere in looking for fortune, and search where I may find ore, until at last I come across it, which will easily retrieve my previous losses.

On Iron Ore, Which Is Found in Certain Countries in Various Shapes

Iron is the foremost among all those metals which God Almighty created immediately in the first age of the world and provided for the use of man.

Iron ore is found in the bowels of the earth in various forms, but it is very difficult to recognize it. Nor will you make it out at once when smelting it in a fire, but only when you have assayed it many times.

The foremost ores are those of Lebanon in Syria; they were mined in many places there even before the Flood, in Gilead and on the rocks of Hermon, where Tubal-cain, the first smith, had his forge.

These ores extended in the Jewish land beyond Mount Carmel, running from the Phoenician border through the land of Cabul to where the Assyrians lived—smiths of the Jewish race. G2

In Colchis, too, where the old Chalybes lived, they had iron ore with a strange property: it produced iron as white as silver when smelted in a strong fire with pyrite.

The *pyromachon* of Aristotle.

In Spain where those famous rivers Chalybs and Bilbilis are and where the smiths claim to be Chalybes, they have ores such that iron resembling steel is wrought.

In Styria also they dig ore with the property of yielding both steel and plain iron; but steel has to be smelted twice, while iron is smelted only once.

It also happens that silver is smelted here from another kind of iron ore, near Goldkranach and everywhere in Styria where pieces or lumps of ore are found with native metal inside.

There is also iron ore containing copper, but it always yields unweldable iron. This rawness may be rectified by bog ore, if this is added and [the two ores] smelted together.

Bog ore will alter any ore and will commonly make the iron weldable, but if too much is added, it will make the iron red-short and thus do it harm.

There is also an ore which has to be broken like stones. There is plenty of it in the Mrzygłód domain in the county of Pilica, where it is called Niwka ore, G2v
from the name of the village in which it was first found.

Hence to this day the forges in this region, where iron is made from this ore, are called Niwka forges; there cannot be in Poland any iron of a more notable or better nature.

[The meanings of dul *and* grąpie *are discussed in the Introduction.]*

This ore is so good that steel can be made from it. A *dul* must first be formed from *grąpie*; this requires charcoal from young pine trees, since charcoal from other kinds of wood won't do.

[A "monk" is probably a lump of iron in a form suggesting a monk in his black habit.]

Niwka ore has the following properties: if it is overheated or heedlessly blown to a "monk," so that the slag will not pass out of the bloom, then the iron will be brittle and very poor.

Various forms of ore are found there. Plenty of it is always being dug out of the earth, but its properties are different in each mine: it may be hard and rich in iron, or crumbly.

For this reason the ores are commonly mixed, and the iron obtained is better than iron from a single ore. This I know because workshops there were given into my charge.

Sometimes nonferrous ores may be found instead of iron ore, but these cannot be won because of the abundance of water [in the mines], or else the power of God preserves those ores, their destined time having not yet come.

There is also an ore there which abounds in pyrite. It is very rich in iron, yet the iron it gives is poor. It will spoil every other ore and will not agree with any; for this reason forgemasters avoid it whenever they can.

G3 There is many a place where ore is sought. It happens sometimes that pyrite is found by itself, having an appearance like fire and giving fire like flint does.

This stone has its own curious way of retaining silver and tin, as well as copper, and there is something even more strange about it, for vitriol and alum are also made of it.

It is of various colors—yellow, black, white—and is found in the earth in large lumps. [On heating] it yields a red pigment and sulphur, and it contains some gold, but the stuff is of no use in the fire [of the smelter].

In the regions of Ogrodzieniec, Olsztyn, and Krzepice there are everywhere stony iron ores like those of the Niwka region, but they are not in quality and merit as excellent as the Niwka ore, and are not so speedily smelted.

Such ore should be roasted on a grate just like silver ore and then washed and beaten with iron flails; all the sand should be removed from it with a sieve.

There is still another ore having different properties. It is found in bogs in ample amounts, lying in the water beneath the vegetation. It is called bog ore, but for its virtues it might be called golden ore.

It is very easily smelted, requiring no roasting, only to be washed free from sand and poured into the furnace. It exceeds every kind of ore in its yield, but gives a very brittle iron.

Another kind is so rapidly smelted that it does not agree with any other ore and does harm to every ore.
Rocky ore is therefore mixed with it, and this reduces G3v
its excessive fluidity.

A third ore of this kind *[hałda]* is yellow, fine, of the same form, greatly resembling the previous ore. It is often met with on lowlands in alder thickets; it gives extraordinarily good iron.

Of such kind is the ore in the Margraviate of Moravia, the part of Ołomuniec next to Silesia, in the domain of Szternberg. The ore there lies so deep that it has to be hoisted by means of winches.

In the Kingdom of Bohemia such ore is also plentiful, but it differs from other ores in that the first blowing will bring but poor iron—it will, however, set tight and hard in the furnace.

You have to smelt it a second time in a strong fire if you want to forge good iron. The forgemasters test this ore by smelting it in the fire twice.

This ore occurs in the lands held in common by the lords of Zeydlic near the boundaries of the Margraviate of Moravia in two places, separated by only half a mile one from another, and has the same properties in both mines.

The nearer ore mines are at Borowa, the further ones above Rąńsko.

A very yellow and very fine ore, unusually good, occurs in the lands of Krakow on the territory of Jeleń. It lies in a dry place on a hill near the river Przemsza, but the floor of this bed is very deep.

There is an ore of the same form in Silesia, in the domain of Bytom. There is no other in Christendom of similar virtue. It is commonly called Piekary and Żyglin ore, from the places where it is dug.

The nearer ores are at the village named Piekary, the further ones at a small town named Żyglińskie Góry.

G4 It lies in the earth at a depth greater than other ores, up to five or even six fathoms deep. It is hoisted to the surface by a winch, for it is so deep that the eye cannot reach from the bottom to the mouth of the shaft.

It is so fusible that it may be charged into the furnace without roasting, like bog ore. Larger pieces are also found, but they are not liked on account of the labor needed to roast and break them.

It gives iron so good that nowadays there is none better in all Europe: weldable, tough, not brittle. Sometimes, when it has been properly treated in the fire, it is stronger than steel.

When it is heated under pine charcoal, its quality is so high that it does not yield to Styrian steel; indeed it is better in that it won't be wasted by breaking, as often happens with Styrian steel.

These are the reasons why the noble iron from this ore always and everywhere outstrips other kinds of iron. The honorable town of Nysa, in Silesia, has long been famous for its excellent stock of the best of it.

Forgemasters made a good profit from these ores as long as the officials closely watched over the mines; nowadays, through lack of supervision, this ore is spoiled because it is always supplied mixed with waste.

The best ore is marred by carelessness, which causes considerable damage. Fallen-in pits are only scraped at the top, so that mere refuse is supplied from them.

Thus the forgemasters suffer damage, and their lord no less, as the bulk of the ore is left in every pit, G4v
covered by the earth that has fallen in. See what damage is done by careless miners!

There is nothing to praise nowadays. There is no vigilance such as there was in olden times when the officials went down into the mines and took proper care of them.

Of the Disappearance of the Bytom Ore and the Emergence of the Tarnowskie Góry Ore

There is within the domain of Bytom a great abundance of iron and silver ores; there is also the pyrite stone from which alum, sulphur, pigment, and also vitriol are obtained.

Because of this ore—before it disappeared—Bytom was once called Silver Bytom and was famous everywhere as such. Having silver enough, the citizens of the town would adorn their bedchambers with silver steps.

For a long time they had great fortune with these ores, as long as they leased the mines from a certain Szarlej; but when they learned that he was an earth spirit or guardian [of subterranean treasure], they came to loathe his company.

In order to drive him off they kept on annoying him in different ways: they wrote spells and made crosses which they laid upon his ore.

Aware of their plot, Szarlej led them to the mines and gave them some news which they were not very glad to hear: that on account of their malice they would lose their ore.

There was a loud noise, and Szarlej ordered all those in the mine to flee to the surface, as water would H
soon flood the mines like a sea. At once a huge wave broke forth, as he had said, and flooded the mines.

For this reason this lake is still called Szarlej. Instead of ore, fish are now mined through all the deeply flooded pits where the rich ores were drowned.

Later they often tried to find ore in other nearby places, but they never had the same fortune as at first; water always hindered them, and hence the mines were abandoned.

Since according to God's word every hidden thing in the world must be made manifest and discovered, the treasures of the earth, the metals and ores, are always revealed to mankind at the proper time.

They disappear in some places and appear in others. Some vanish, yet others are revealed. They commonly come to light from the secret places of the earth, by chance.

Thus by chance, after the silver ore at Bytom had disappeared, there soon appeared in the mines an ore different from that of the flooded pits a mile away. It was found by a certain peasant from Tarnowice named Rybka.

A hundred years earlier nothing was there but a thick forest from which the surrounding villages took their names. This forest was composed of hazel trees, blackthorn thickets, and great pines.

So Lasowice was named from the forest *[las]*, next Tarnowice was named from the blackthorn fruit
Hv *[tarnina]*, Sowice from the cry of owls *[Sowa]*, and Leszczy from the hazel tree *[leszczyna]*. In this way were these hamlets named from their surroundings.

Amongst these, almost in the center, Blaszyn was called after the lord of this land, whose name was Blacha, but this was only after the forest on this place had been cleared and a town had been built.

Near this Blaszyn one day the aforesaid Rybka, looking for lost cattle, found, under an uprooted tree, ore that had been wrenched up to the surface of the earth with the roots.

Later, when people learned of this ore, they started digging in that place with great greed. When they reached the bottom, they found a large amount of ore.

Others, seeing this good fortune, joined the diggers, and soon there gathered a greater company. They dug up the earth and turned it over, sank shafts, finding everywhere a great abundance of ore.

This first mine was later named Świeca ["candle"], which meant that after having sunk the shaft, the miners set out in various directions with candles and found rich ore in plenty everywhere.

After the year of our Lord 1526 a statute was passed, at Bobrowniki first, concerning the place where the ore had been found.

At that time the Margrave Jerzy, Duke of Brandenburg, and with him Duke Janusz of Opole, brought these mines into a proper state and introduced excellent order.

There were scarcely three miserable huts there at the time. Many people dwelt under the trees in shacks covered with turf, as they cared naught for a pleasant abode once they had found the ore. H2

The Duke then ordered that the forest be felled and a town built instead. The people who had come together from all quarters speedily erected, while mining the ore, a considerable town.

It was named Tarnowskie Góry, from Tarnowice, because a great plot of ground was torn off from this village and a large part of the town built there.

It is a free town, with many liberties, much trade, and many markets. Any citizen may bring any goods he likes there, and nobody will prevent or hinder him.

Great markets are held on Saturdays at which salt, iron, and corn are sold. Rich and poor will get everything they want according to their means.

The town, abundant in every form of wealth, excels all Silesian border towns. Its inhabitants are benevolent, simple, not haughty, and especially clever in carrying on trade.

On the Ores and Precious Stones Found in the Earth in Certain Places in the Duchy of Silesia

H2v Our Lord in his grace gave to the land of Silesia great underground wealth in certain places. It has ore of such quality that the iron smelted therefrom has no peer in Europe.

There is silver ore enough, and copper ore too, which is found in the Ryphaean Mountains not far from the river Bober, near the ancient borough of Kupferberg [Miedzianka], which takes its name from copper.

Sapphires occur near Goldberg [Złotoryja], Leoberg [Lwówek], Bolesław, and Hirzberk [Jelenia Góra]. Diamonds are found near the castle of Kynast [Chojnik] in the fields and in pools through which the streams from the mountains run, in the meadow of the Izera near Gryphiberg, in Gusmansdorff, and in many other places in the Ryphaean Mountains.

There are ores there, and in certain rivers gold dust and various precious stones are found: sapphires and diamonds of extreme beauty, in shape and brilliance like those from India.

Curious jaspers of various colors are found in Irzkienberg among the Sudeten Mountains; jacinths and rubies are found in the river Izera as well as in the pools from which gold dust is extracted.

At Hirzbrun [Cieplice], at Kupferberg, and at the source of the river Elba [Laba] loadstones have long been taken. Near Gryphiberg [Gryfów Śląski] beautiful pearls as large as peas are now extracted from the river Kwisa.

Margaritae fluviatiles [river pearls].

The Izerwiese, the meadow of the Izera, is a place on the borders of Silesia in the western corner above the rivers Quisum and Lusa.

Garnets—glittering little red stones—are found in Izerwiese near the river Izera and in the Sudeten streams. Near the great Riesenberg [Karkonosze] Mountains excellent amethyst is found.

Topaz is found in the Sudeten Mountains near the larger lake of the Riesenberg and in the fields of Szyłdawa [Wojanów]. Turquoise, blue and tawny in color, is found in Izerwiese near Strzegom and in mountain streams.

Furthermore *terra sigillata* has appeared in this region, a substance worth more than gold, for it has the wondrously strange property of counteracting venoms,
H3 poisons, and every disease.

Experience shows it to be better than the Lemnian variety, which is extracted from old mine shafts at great expense and carried overseas from the isle of Lemnos, and which has sometimes proved to be greatly adulterated.

In the duchy of Świdnica, near Ostra Góra in the land of Strzegom, this precious earth is engendered in old mines and pits where gold ore had long ago been dug or won.

This Silesian *terra sigillata* was discovered by the research and industry of the most renowned Doctor Johannes Montanus.

When it has been taken from old pit shafts, small round cakes like little plates are immediately prepared from it under official supervision and are marked by stamping with the town seal.

Fiery Exhalations, or the Emissions of Fume from Ores, When Fire Bursts Forth from the Earth at Night and Reeks in Daytime

Miners who dig ores are often guided to such treasures by exhalations, which twice a year, in a marvellous manner, send forth lights as bright as fire.

When appearing at night, this exhalation has always a flamelike form or color; but in the day time there is no flame: it only rises from the earth like mist or smoke with a great hissing.

Fearsome it is to go where it appears, because the earth opens up at that place and everything that is on H3v
the surface immediately falls into the depths.

Because of this, miners often find in their ore, particularly in dense forests, bones, leaves, chips of wood, and burned branches which the earth had swallowed and scorched by an exhalation.

In the mountains near the Sudeten range, people living among forests where there are ores often see such exhalations at night in spring and summertime.

These exhalations are like the will-o'-the-wisps that appear at night in forests and sometimes lead wanderers and travelers astray.

The Germans commonly call these fires flying about at night *Irrwisch* [mischievous imps]. In reality it is the devil who whirls in the air to make sport of men in his fiery shape.

The Polish people call them *latawce,* from *latanie* [flying] ; witches keep them for their spells to bring them money and corn, for they are masters in theft.

On Ghosts or Mountain Goblins

Virunculi montani [little men of the mountains] .

There are also strange goblins, who commonly live in the earth amidst the ores. They keep company with ore-miners, as if belonging to a common guild. They dwell in those places where there is the most ore.

H4 They appear underground in various shapes, rambling in the galleries between the pit shafts. Sometimes at night they walk on the surface over places where ores will appear.

If one of these elves appears in miner's garb with a pick and raps around the pit, the miners at once gladly dig the place, for there they will find ore in abundance.

These elves do no harm to miners in the pits, but come to them with cheerful smiles. Sometimes when the miners watch them, it seems that they are also digging,

but no sign remains of their work when they disappear from the spot. Yet if someone reviles one of them or mocks him, the elves will soon take revenge for the offence.

This once happened at Bytom, when miners played spiteful tricks on a demon called Szarlej. He flooded all the mines, and the marks of this still remain.

The Germans call such peaceful man-imitating goblins *Kobolds,* and some call them underground guardians because they willingly stay among the subterranean treasures.

One jocose ghost among them used to appear near Riesenberg and play tricks on men. Notwithstanding that he appeared to people in a terrible aspect, he usually did no one any harm.

But if someone mocked him or did him mischief, he
returned the insult in this way: he dimmed the man's H4v
sight and led him into a marsh, which caused the man
much trouble before he got out;

or else he drenched the man with water and led him through marshes, so that the man did not know where he was going all day long. Sometimes in bright daylight he would dim the road with dark clouds and raise a downpour with great thunder.

Another kind of goblin, cruel and inhuman, is mischievous and antagonistic to miners digging. A custom they have, when rambling about in the pits, is to amuse themselves by overwhelming the miners with fallen earth.

A goblin of this kind used to haunt a pit at Annaberg; it appeared with a great long neck like a horse, so long that in the end he strangled in his satanic wrath twelve miners in that pit.

In our country a goblin of this kind sometimes appears at Olkusz and in some places does much harm to the miners when they are digging unthinkingly; hence many of them perish there miserably.

Philosophers say that súch goblins are dumb; and they lack the power of reasoning. They always come to people in a fearsome shape, and they are quick to do harm to drunkards.

Dwarfs or Demons of the Forge

In olden times some elves of this kind, but small as children, used to haunt forges. They readily came to forges on Saturday night after the smith had finished his work, when the fire had been put out.

When the smiths had quieted down and gone to sleep, I
then the dwarfs would forge the whole night through.
When the cock crowed or dawn began to break, they
vanished and the hammering stopped.

At the forges of Jędrysek and Kot strange things happened for a long time because they were haunted by such elves. Every Saturday evening food was placed by the small furnace where they used to forge.

Forgemasters commonly called the creatures dwarfs and greatly valued them, thinking them holy. Hence they never did them any mischief when they were forging, but held them in respect.

Those forgemasters whose forges they haunted had great luck, everything went well for them, almost according to what they had in mind, both in the craft of ironwork and in everything else.

When the aforesaid forgemaster Kot saw in his forge these elves forging barefooted, clad only in shirts, out of piety and touched by pity he gave them clothes in reward for their toil, but

when the clothes were placed where they used to forge, they immediately left the forge complaining and saying: "Our master is paying us off and cares no more for us, so we shall leave the forge and not delay our going!"

These apparitions and dwarfs were thus scared away from Kot's forge, and soon afterwards Kot's luck in his forge changed for the worse.

Iv

The Proper Organization of a Forge: The Construction and Maintenance of Forges, with the Arts of Charcoal Burning and the Blowing and Forging of Iron

If you do not know how to organize a forge, or how to manage your workmen, then I advise you to give up your forges and think of other ways to earn your bread.

This management, believe me, is a difficult knot, which it is not given to everyone to unravel with ease; it is hard to understand. You won't learn everything about the organization of a forge in a single day.

You won't comprehend the work in a forge all at once, or how to manage an ironworks; there is, for one thing, the maintenance, and for another the techniques for blowing a sound bloom and forging the iron. It is difficult to understand straight off how to deal with all this.

There are five arts which every forgemaster should

study if his forge is to be properly run: first he needs knowledge of the forge; he should himself know how to do all that his craftsmen do.

Next comes abundance [of supplies] ; this is always needed. Without this, you will not earn a loaf of bread. You must always have plenty of charcoal and ore if you want blowing and forging to run smoothly.

Foresight is also necessary, and you must be guided
by it in organizing a forge if you don't want to go
astray. You must persistently look for benefit every- 12
where and always beware of damage.

Close attendance and assiduous supervision day and night are necessary for this purpose, for if you are a negligent master in this respect, you will soon fall into misery and do away with the forge.

Above all, you must have the forge on a dependable watercourse, which should have its head not low but high, so that the waterwheels do not wade. The river must be deep and wide, lined with planks on both sides.

The smelting house must be spacious, covered against rain and duly closed in on all sides. All the machinery should be in good order, and all the sluices and channels, piles, slotted wall-pillars, footing-beams, and posts should be sound.

The workmen's huts should be properly built and set up near the smelting house. The ironmaster's house should also be close by the smelting house so that he can overlook all the forge.

The bloomery hearths should not be close one to another, neither should they be on a marshy spot where there are springs, for the sole of such a hearth would become moist, forever causing the iron to be scantily heated.

The hearths should be smooth, built of good clay, not very high, but low, since a high hearth would burn a lot of charcoal and cause the blast to be slow and tardy.

The chimneys are to be properly shaped, spacious,

and raised high so as to prevent sparks from falling on the bellows and setting the smelting house on fire.

12v Now let us turn to the hammer, where the iron is being forged. It is necessary for you to know how the machinery stands there: whether the smith's hearth is properly constructed, whether it is in good order and sufficiently spacious;

whether the tuyere is sound and appropriately set in place, with an opening such as will cause only little charcoal to burn and bring about the formation of fairly large *duls,* their iron whole and sound;

whether the trough near the smith's hearth always contains fatty clay, well mixed with water, for spreading on top of the glowing charcoal in the hearth for better heating.

The hammer should be large enough, neatly formed, not too high, smooth on all sides, and should have a smooth, hard face, so that it will always forge a bar evenly.

And you must watch that the hammer runs evenly, so as to strike the same spot every time. The waterwheel should be capacious and the shaft thick, running fast and with great driving force.

[The names of some of the particular tools used in a forge are probably not well understood, since the same name did not mean the same implement in all localities and periods. A listing of the Polish terms can be found in the glossary on pp. 114-115.]

The [iron] sleeve of the hammer helve must be thick and tough and the bushings should be faced with steel and set squarely into the hammer uprights. The recoil beam should also be thick and long enough, and the cams must raise the hammer uniformly.

The anvil, which is struck violently from above by the hammer as it continually forges the iron, should be smooth, undamaged, properly faced with steel, and firmly set into a thick stub of oak.

You must have good stores of tools for forging and
13 bloomery blowing: namely, tongs, a piercer, a hot set, crowbars, swages, a tuyere hook, and shears which should be steeled.

Always give heed to repairs in the smelting house and to the anvil and the hammer, and take care of the bellows, as they cost a lot; you must look everywhere to see that the maintenance is proper, that there is no damage anywhere.

Be careful to check frequently that the wheels are running evenly; that the bearings are not wedged more on one side than on the other; that the bellows are blowing properly; that their leathers are sound, not torn and without any holes;

that the valve flaps are not worn; that the bellow boards and the bellow cams are sound, as well as all the driving mechanism; that the bellows run evenly and smoothly, without any rattling; and that the driving mechanism does not run screechingly.

When the bellows are new, make them run smoothly, for they will always run as you make them go at the outset. You must watch that the bloomery-smiths don't let the slag enter the bellows, which would then soon burn away.

These mechanisms for forging and blowing are in constant need of tallow for greasing the bearings and cams, since, just as a cart runs faster when greased, so the blowing and forging go faster with greasing.

Bloomery-Smiths or Smelters

You should secure in good time skillful bloomery-smiths whom you should properly respect, not withholding the pay due for their toil on Saturday of every week.

Besides their wages, they also should receive bounties 13v
according to their rank: first the elder, next the younger. Likewise the smiths and the charcoal handler, but the wages of the maintenance master should always be greater.

For maintenance a good bloomery-smith is needed, skilled to undertake this task, knowing how to repair a bloomery hearth properly, how to make a proper opening in the tuyere and how to insert this into the hearth.

You will always have great profit from a forge when you employ skilled, good craftsmen. Beware of unskillful ones, don't keep such trash if you don't want to come to misery because of their poor work.

The maintenance of a bloomery should be properly understood: how to set the tuyere in a hearth and how to shape the tuyere opening so as to make it heat the hearth well without burning too much charcoal.

For when a long time is taken in blowing a bloom, the profit decreases because charcoal is wasted. The charcoal in the hearth should be frequently watered, because it costs a lot and so should be conserved.

Every bloomery-smith who is expert in his job should have his hearth well preheated before putting on the blast, for when the hearth is warm, the blast speeds better, and in this way but little charcoal is used.

Whereas on a cold hearth, it is seldom that anything can be smelted out; slag alone will be mostly burned, which makes the iron dungy, filthy, hence bad and hard to work.

You should keep an eye on the bloomery-smiths from every side, for these descendants of the Cyclopes are
14 so trained that they are not ashamed to poach a piece of a bloom and then throw it into the water [to hide it] when nobody is looking.

Or else at night, he will slip the first iron out of the hearth and, having put it in the water, will rush with it to his hut. When giving it to the smiths he says it is a *grąp* that had formed beneath the bloom during the blowing.

Such iron-snatchers have always angered me, and I would be unwilling to keep such villains, since the iron will often set badly in the hearth and much charcoal will burn when the blast continues so long.

He could blow two blooms out of the charcoal and ore that he uses for one, and yet his bloom will be poor. He will have blown only *grąpie* at the sole, while at times he will blow something like a stub. All this will cause fourfold damage,

first because of the ore, second because of the

charcoal (for much will burn), third because the extraction will be slow, and fourth because he will not always get a bloom, not even a poor one; it would be very lucky if he smelts out anything at all.

The hearth must be looked after: the ore must be evenly charged into it and the charcoal should be properly cleansed of sand, since the iron is bad when the ore has been charged improperly or the charcoal is mixed with sand.

Too much slag should not be kept in the hearth since the bellows would not be able to blow through it and the slag would flow into them, causing a chilling of the iron in the hearth.

A careless bloomery-smith will soon cause damage, especially if he is a drunkard or sluggish; he will soon burn the bloom if he does not look after the hearth, and he will damage the bellows by running them foolishly.

Blacksmiths 14v

The forging of iron calls for skilled blacksmiths who know how to hammer iron properly. Engage them early and never keep idle or clumsy fellows for smith's work.

If you cannot engage blacksmiths skilled in the practice of their craft, then give up the working of your forge until you can find good smiths somewhere.

Bad blacksmiths will soon bring you misery: you will lose the money laid out on the forge, you will also lose the profit and toil in vain. This has happened to me through having bad blacksmiths.

There are men who do not know how to repair [the plant] or how to forge properly, yet undertake these tasks. Later such a braggart worker will prove to be an excellent spoiler and damager.

Notwithstanding his repairs, his constant knocking the hammer about, the work will never go smoothly, he will not set the hammer properly, nor the bearings and shafts. Everything will go wrong.

The recoil beam being wrong, the bushings too low, and the sleeve of the hammer helve badly set, then the tup falls unevenly over the arms. This will break every bar, spoil a lot of iron, and bruise and break or knock off a part of the anvil or of the tup itself.

When heating the iron such a man will burn so much that it will give you a headache just to watch. Isn't it
K a pity if by his slow hammering a lot of charcoal is wasted?

Because of all this, every forgemaster should have a master well skilled in the art of smithing who knows how to maintain the hammer so that the work goes smoothly without mishap.

After having established such order in your forge that none but good craftsmen work there, you must have an overseer acquainted with the forgemaster's customs.

If instead there is someone who does not understand a forge and does not know how to deal with the smiths, the forge will not run its proper course, and so its profit will be lost.

The overseer must be faithful; he must always watch the smiths while they are forging so that they do no damage. If he is heedless, he may be robbed before his very eyes, and the foolish warden will not notice it.

A watch should also be kept when iron is being heated, as the cunning fellows know how to hide it then. They know how to push back a piece of bloom and say that it has burned, and all the time it lies in the fine charcoal dust below.

They will return and hammer it late at night when the overseer is not watching. This is the tip they give themselves; they think it a harvest when they steal something for beer.

They should also be watched on every side when they are working iron into bars. If the overseer is not on the lookout, the smith, with a flick of the shears, will cast a piece of bloom into the water between the wheels.

The charcoal handler will also sometimes throw a bar into the clay slurry stirred up with water, or else he will throw it under the wheels or into the troughs. Who can reckon all their artifices! Kv

Sometimes he will hide a piece of iron in slime and will take it away when no one is looking. If he has no chance to filch anything, then for spite he will burn the iron in the furnace.

I speak of the malicious workmen. There is no harm in the good ones who always work honestly, as they should. Everyone must work faithfully and always beware that he never does harm to his master.

The Charcoal Handler

You must have an agile man as charcoal handler to serve the small smithy hearth; he should always keep the hammer clean and watch the hearth when it is running.

Before hammering the iron, he should first clean the small hearth so that it contains no slag or loose cinder; next he should put clean charcoal dust in the bottom of the hearth.

Then he should put into the hearth crushed charcoal, well cleansed of sand, gravel, and stone. Only then let him start the bellows and pour water mixed with clay onto the fire; let the iron get heated.

If he often pours such water onto the glowing charcoal, the iron will heat up quicker, and less charcoal will be burned, especially if it is weak, for this hinders spurting and cools it at the top.

When he uses uncleansed charcoal, the heating of the iron will be bad and slow. It will cool down in the furnace and a *dul* of iron will form there that is impure, cold, and imperfect. K2

The charcoal handler should carefully gather the *grąpie,* which sometimes break off from blooms when they are shingled or from iron when it is forged. He should inspect these *grąpie* and throw them into the fire in the hearth to be heated into a *dul.*

Every time the hearth is blown out he has to count all the smith's tongs and all the tools and implements for ironmaking and put them away. He must also see that they are sound.

He must also watch over the millpond, so that an excess of water does no damage to the sluice or the smelting house. In time of drought he must stop the chinks of the hatches all around and not let the water run out without being used.

He must keep stock of the charcoal in the smelting house. Along with the bloomery-smith he must turn over every basket of charcoal brought in to see that there is no defect in it. He must also count the number of cartloads each pile yields.

Later, when making his reckoning with the charcoal burners, the charcoal handler must hand over the tallies he has made of the charcoal carts, showing how much each charcoal burner has worked.

Day and night, he must, moreover, watch the bloomery work in the smelting house when it is running. Each time, working with the bloomery-smiths, he must draw out the blooms under the hammer and shingle them.

When the smiths are forging, he must obey them in everything, doing swiftly what they order him to do and doing nothing on his own, because in such work one must be quick. This is how the charcoal handler must behave in a smelting house.

K2v

The Charcoal Burners

You must give money promptly to the charcoal burners for their work, so as to let them buy bread, for, believe me, no one will be so stupid as to work in a forest if he is hungry.

First, enough piles must be prepared for winter; money should not be spared for this. During winter you might let wood be hewed by the cord so that in spring you can pay off your last year's debts.

You might detach the charcoal burners for such work

by the cord on Saturdays, but you must pay them at once; they will then work more willingly.

You must see that they split up the wood small and chop the knots out of the branches. A forgemaster can make great profit from cordwood, and he will also obtain charcoal more easily by having the wood available.

For this purpose plenty of young growth should be cut each spring; it will give a clear profit in [the following] winter since you will get charcoal at little expense so that your forge will run without ceasing.

Every charcoal burner must be diligent at his work. He must go into the forest early and work until night. He should plan his work in the forest so that he can prepare two piles annually.

One pile must be covered with turf well before winter. It should be set up on good sand in a dry place and covered with sand, so that water from melting snow is not absorbed in the winter.

The other pile must be ready early in spring. It can K3
be prepared during a winter thaw, since winter is as good as summer for charcoal burning as long as the wood is not frozen.

Order should be observed in this work, in due accord with the ancient customs of the old charcoal burners.

First, when dry logs have been provided for the piles, one man should help another in setting them up. Thus one day there is work at the pile of one charcoal burner, and the next day at another's.

When set up, the piles should be promptly covered with turf against the winter, and thoroughly secured against bad weather. The piles should be burned in the same sequence in which they were set up.

At each great charcoal burning every man should willingly help the others, up to four days and four nights, until the fire reaches the logs at the bottom.

The man to whom the pile belongs should provide for this burning an eighth of a tun of beer with which he

[A tun is equivalent to 27 gallons.]

should at that time treat his companions, and they should act as follows:

A master skilled in the art of burning shall be appointed for setting up a pile. He should know how to direct the fire within a pile properly. He should, moreover, be able to endure the smoke and dust.

In Germany, charcoal burners carry out their task thus: instead of turf they cover the pile with brush-
K3v wood. When they have no brushwood, especially that of fir, they use green grass for this.

Next they place wet sand on the brushwood and put in the fire at the very bottom. In this manner the fire in the pile will advance slowly and the charcoal will be keen and good.

In our country the piles are covered with turf, and the fire is put in from the top through a vertical vent in the middle; then sand is put on to check the fire, and holes are pierced in the turf to let out the smoke.

At the beginning, shavings and chips are always prepared before fire is put into the vent. Next they are burned until the fire spreads all over the branches in the pile.

This manner of charcoal burning needs, I think, more toil and is more difficult, since in this country ten men can hardly manage the task, while in Germany a charcoal burner can do the job with but one other man.

The fire in a pile must always be kept slow, and if you don't want to get ashes instead of charcoal, you should put plenty of sand on that side where the wind might blow through the turf to the fire.

If a strong fire breaks through the side, you must immediately block up the hole with sand. When the wood is dry, it may easily crackle so hard that it bursts out away from the pile.

At such a time no one should sleep, but all together should cover those places with turf; everyone must promptly bring sand and when the fire breaks out damp it down with layers of this.

Take care that the fire in the pile does not gain K4
strength at night and does not wax too strong; to do this you must spread sand all over the pile from the top and tamp it with a flattened pole. Do this continually.

The fire in the pile should not be damped down too much, for this might harm the charcoal; rather, let the fire run slow and steady, and see that the smoke is blue everywhere.

When the smoke is dull, then the charcoal is stumpy, slack, and poor to work; but when the fire burns slowly with blue smoke, then the charcoal will be keen and good.

You must yourself learn how to do charcoal burning, so that you may know how to behave with these Cyclopes; you must understand their customs if you are to rule them.

There are skillful dodgers among them, always ready to cheat their master. Such a one may appear as humble as a Bernardine monk, but within be a downright rogue and clever shuffler.

He would gladly take the risk if he got the chance. The slacker sets at naught the work in the forest. Hence he must be always kept in check and watched; don't give him a chance to idle—let him work with others.

I would not advise you to give earnest money for the work of a rascal who has learned to be idle; when he is in debt, he will skip, and to hell with the debt, leaving some old useless billets in the forest.

In this guild there are also those who scamper from one forge to another and do not stay very long any
where. Work comes hard to a man like this; hence he K4v
has many a master every year.

There are other rascals who have a habit of making bad charcoal on purpose—very thick—because a basket holds less of such thick charcoal than of fine.

These thieving charcoal burners deserve the gallows for such work, or at least to be driven away from the forge. This would be an example for others not to make such base charcoal,

which does not heat the furnace, but simply burns away like dried wood. This causes much damage in the forge and much loss: all the profit and gain of your toil will vanish.

Again there will be a notable loss if a half-filled basket of such charcoal is brought to the smelting house: if coarse charcoal is heaped up in the basket, it will not be full when brought to the smelting house.

There is a special penalty for such villains, established long ago by the old ironmasters, who used to punish the rascals by counting two baskets as one if not properly filled.

The basket should be filled so that it arrives at the smelting house with charcoal up to the level of the hoop. It should not be heaped up in the basket so as to seem much, but should be solidly packed into the basket.

You should always beware of such a smoke-stifler. Let him have no place among the workmen, so that others will not follow his example of roguery and do such work as he.

Perſeus exultat dum conficit enſe Meduſam,
Atq; premit celeris tergora victor equi:
Semina nequitiæ quiſquis fugit, ille beatus
Laudis inoblitæ præmia Victor habet.
Et quoniam rigidi firmo tutamine ferri
Vicit ouans: ferrum eſt res precioſa nimis.

Perseus springs forward, having killed Medusa with his sword, and victoriously advances on the back of his swift horse. Like one who has fled from the seeds of wickedness, he, a blessed victor, has conquered triumphantly with the firm protection of stiff iron and so merits the reward of never-to-be-forgotten praise. Iron is a thing of the very greatest value.

Lv On Iron, Which Surpasses Every Metal in the World by the Power of Its Virtue and Excellence

Iron is by its properties the strongest of metals and the most necessary to man. Its sharp strength in battles has caused the downfall of great monarchies in the world.

Its sharpness will destroy everything in the world; it will cut through the greatest tree and cleave a rock. Terrible is its sharpness, with which a furious marauder may inflict incurable wounds in a fight.

It brings shameful death to unworthy malefactors, and immortal glory to the worthy who use it in a knightly manner in the defense of their native land from the infidel.

Hence noble people esteem iron for its great virtue above all treasures and jewels; they value it more highly, especially since they can use it for purposes of war.

In times gone by, great monarchs ruling the world all loved it; with iron the king of Macedonia conquered the world and made all potentates his vassals.

Who can appear in battle without armament made of
iron? Are silver armor and a sword of pure gold
L2 worth anything? If you haven't good iron, it will be
hard to have trial of arms with bloodthirsty Mars.

A battle-tested sword of good iron hung at the side is a noteworthy ornament. Many a man has won great glory by it and has acquired a coat of arms, the sign of nobility.

Perseus acquired immortal fame because he beheaded with his sword the ruthless Medusa, who, it is said, was so murderous that she killed men with her terrible glance.

Mercury also used this weapon and with it killed the hundred-eyed Argus. It was with this weapon that Jason won the golden fleece in Colchis and carried Medea off from there.

With his iron sword Hercules terrified all nations and won his immortal glory; wherever he turned, it caused fierce beasts and cruel men to fall.

Hector, the great conqueror and defender of all Phrygia, having left Troy, perished miserably by iron, and Achilles, who seemed to be invincible, also lost his life by iron at the gate of Troy.

By iron was Ilium or Troy destroyed and a great many of its people killed. There miserably fell the house of Priam, with whom the Greeks contended in war to avenge their wrong.

Babylon fell by iron and likewise Carthage, where there were a hundred thousand armed men with whom sly Hannibal contended against the Romans, fighting a long time with them for his liberty.

It was by iron that the Romans long troubled the world and ruled it and all its kingdoms. He who dared oppose them came to destruction; with their iron they were fierce to everyone. L2v

The Romans so loved iron that they wore iron rings on their fingers in preference to gold. This they did to show that it was by iron they had subjugated the world to their power.

This custom lasted a long time. Moreover, an iron crown was made in Rome which was worn as a sign of dignity, and some of the Caesars were crowned with it.

Od [i.e., Og], king of Bashan, a great pagan man of the race of giants, had—so Moses writes—a bed of pure iron which he had ordered to be made as a mark of his rank. Deuteronomy 3.

German princes formerly had such a high regard for iron that they minted iron money; this money was in use for a long time and was accepted in both Germany and Italy.

Many an honorable town had the same custom and followed these princes in minting iron money, which they used instead of silver coins.

Now let us turn to husbandry, which needs iron tools, implements, and vessels. Axes are needed for wood, and scythes for cutting chaff and grass to feed the cattle.

L3 For tilling the soil you need a ploughshare and a coulter, such as peasants use to plough the earth for their bread. You also need a ploughshare to turn over fallow land.

Moreover, men need iron harrows for harrowing the soil after ploughing it. They also need mattocks for rooting out stumps and digging up stones when they are present in the soil.

Everyone must bind his new cart with iron. Horses must be shod with iron shoes. There can be no building without iron, since artisans must do everything with it.

Sirach 39.

Hence Sirach, when praising God for all creation, mentions only the excellence of iron. He is silent on other metals for the same reason—that of all metals iron is the most needful.

No man in the world can do without it and manage his husbandry, which God provided for the feeding of the human race immediately following Adam's expulsion from Paradise.

It seems that the noble craft of ironwork was discovered by Adam himself and that he used it in his husbandry and with it gained all the fruits of the earth.

There must undoubtedly have been a mattock for digging the soil, and an axe for hewing wood. With what were clothes made to cover nakedness? They must have been made with an iron implement.

L3v When cattle were brought for sacrifice, was it not with iron that they were slaughtered and cut up? And so even in the first age iron was the beginning of every kind of husbandry.

Owing to such dignity iron is always worth a greater price and value than any precious treasure, since however rude it may be, by its strength it surpasses gold, silver, and gems, which are useful only for adornment.

The Proper Model or Image of the Life of a Forgemaster L4

I don't know why you wonder so much at my person. Don't you like my shape? I am said to be as black as a sweep! I am a smith, and the shape that you see is my own.

It is from fire and smoke that all my body has blackened so much, as you see! I care not for fine and marvellous attire since it would not suit our toil. L4v

I am black like a negro all over my body and am cruelly burned by fire. Just step in to see us, and watch how everyone must continually toil with fire in our smelting house!

Look how my body is broiling! Is it little pain that we have in this fire? The flame and sparks fly into the eyes like ashes. I am constantly roasting in the fire on every side!

If your shirt catches fire, part of your body is burned, and sometimes it won't heal for half a year. I am half-deafened by the crash of the hammers! I must perpetually toil under such wretched conditions.

Our life is miserable. When forging, we must beware of injury at all times. We must take heed, for the hammer has no pity; he who forges carelessly will soon get a bruise!

Even Vulcan himself, though a skilled master in his craft, was nevertheless lamed in one leg. Some say it was his mother Juno herself who crippled him in his childhood,

but I do not believe Juno to have been the cause of his injury. It is more likely he was lamed by a blow of the hammer when he was busy with the craft of ironwork on the isle of Lemnos.

M This is why so many of us are deaf and lame, and I doubt if there is one of us who has a whole body. Look how my skin has stuck to the bones from the great heat of the fire!

Gasping for breath, we work ceaselessly at the fire both day and night, until we faint from the great heat! It harasses us so much that a worker may be hardly alive after toiling in the heat.

If someone who has not seen us before comes to us, he will be much surprised at our appearance. We go about like burned-out torches with but little hair on our heads and chins.

The fire has burned everything, so we care little for adornment. If we have beer and brandy, we have no gloomy thoughts. Since Bacchus will soon knock care out of our heads with a tankard, no one worries.

He is kith and kin to us, for it seems that he and our Vulcan were brothers since Jove begot them both, though the mothers that bore them were different.

He is therefore one of us, and he always passes us a cool drink when we are scorched by the heat. Many a man would hasten to the Elysian fields if Bacchus did not comfort him with a tankard!

In this way we quench the fever we get from Vulcan's fire, for it so parches us that we are sick of soul, and
Mv at such times we must pour beer upon our torrid hearts and refresh our faint spirits. So whenever we have silver or gold, we put aside such mammon for this purpose.

Against charcoal dust, which fouls the throat, we use

brandy, which restores us at once; it also soothes the heavy distemper of a sick stomach and speeds the digestion of any unwholesome food in the guts.

We save nothing for the morrow; all we earn we spend at once. We care nothing for riches, but only for what we eat and drink.

We live in miserable huts, since we do not care for splendid buildings or delightful resting places. Our bed is charcoal dust; it is dry, and neither fleas nor flies annoy us when we lie in it.

So don't ask in our smelting house about the wealth that others long for in the world; you see how painfully we labor and yet we suffer want and misery.

Although in misery, we nevertheless enjoy the liberty which we have from our forefathers. It is our liberty alone in which we rejoice, as this is the greatest reward for our misery—not treasure and not money!

From the age of the Cyclopes we have never been the slaves of any tyrant. Everywhere we are free to come and go; after a year in one place, we can go to another.

We have no written laws or statutes, and we govern M2
ourselves simply by our Cyclopean customs. Having no sworn councillors, bailiff, or mayor, we have among us only an elder master

whom we all hold in such esteem that we obey him like a father. He provides us with our necessities of life: money, bread, and clothing.

BIBLIOGRAPHY

Compiled by Jerzy Piaskowski

a. Printed Editions of the Officina Ferraria

1612
Officina Ferraria, ábo Hutá y Wárstát z Kuźniámi Szláchetnego Dzieła Żeláznego, Przez Wálentego Roździeńskiego . . . W Krákowie, w drukárni Symoná Kępminiego, roku 1612. The book consists of 46 leaves, 18 x 13 cm in size. See p. 34 for a reproduction of the title page.

1933
Walenty Roździeński, *Officina Ferraria* . . . , Poemat z roku 1612 z unikatu bibljoteki Kapituły Gnieźnieńskiej wydał Roman Pollak, Poznań, 1933. 4°, 59 pp., 4 plates. This edition, "published by Roman Pollak from the unique copy of the Gniezno Chapter Library," is incomplete. It does not include Murovius's dedicatory verse or the poem's initial verses; the text of the poem begins with the last two lines on p. Ev of the 1612 edition and omits the marginalia, legends to the woodcuts, headlines, etc.

1936
Walenty Roździeński, *Officina Ferraria* . . . (1612), z unikatu bibljoteki Kapitulnej w Gnieźnie wydał, wstępem i przypisami zaopatrzył Roman Pollak, Katowice, 1936. Bibljoteka Pisarzy Śląskich IV. 8°, XXVI + 113 pp., 9 plates. "Published from the unique copy of the Gniezno Chapter Library with an introduction and annotations by Roman Pollak," this full text edition also contains a commentary based on studies of the text.

1948
Walenty Roździeński, *Officina Ferraria* . . . (1612), z unikatu biblioteki Kapitulnej w Gnieźnie wydał, wstępem i przypisami zaopatrzył Roman Pollak, Katowice-Wrocław, 1948. Biblioteka Pisarzy Śląskich, Seria II, 2. 8°, XXVI + 110 pp., 9 plates. This is a reprint of the 1936 edition, with the addition of a small map prepared by B. Zakrzewski showing the forges mentioned by Roździeński.

1962
Walenty Roździeński, *Officina Ferraria* . . . Oprocowali Roman Pollak, Mieczysław Radwan i Stanisław Rospond, Wrocław-Warszawa-Kraków, 1962. Biblioteka Pisarzy Śląskich, Seria A (Staroślaska), Nr. 1. 8°, XCIV + 318 pp., 9 plates, and one map. This is the complete critical edition, containing a facsimile photolithographic reproduction of the entire 1612 edition, with Polish transcription on facing pages, and extensive annotation and indexes. The present English translation is based upon it.

b. Writings on Iron and Steel Published before 1612

1. Agricola, G. *Bermanus sive de Re Metallica.* Basel, 1530.

English translation in Wolfgang Paul, *Mining Lore* (Portland, Oregon, 1955).

2. Agricola, G. *De Ortu et Causis Subterraneorum. De Natura Eorum Quae Effluunt ex Terra. De Natura Fossilium. De Veteribus et Novis Metallis. . . .* Basel, 1546. English translation of *De Natura Fossilium* by M. C. and J. A. Bandy (New York, 1955).

3. Agricola, G. *De Animantibus Subterraneis.* Basel, 1549.

4. Agricola, G. *De Re Metallica.* Basel, 1556. English translation by H. C. and L. H. Hoover (London, 1912).

5. Albinus, P. *Commentarius Novus de Mysnia oder Neue Meysnische Chronica.* Wüttenberg, 1580.

6. Albinus, P. *Meissnische Berg-Chronica.* Dresden, 1590.

7. Biringuccio, V. *De la Pirotechnia.* Venice, 1540. English translation by C. S. Smith and M. T. Gnudi, *The Pirotechnia of Vannoccio Biringuccio* (New York, 1942; paperback edition, Cambridge, Mass., 1966).

8. Borbonius, N. "Ferraria." In *Nugae* (Paris, 1533). French translation by A. Dufrénoy, *Annales de Mines,* vol. 3 (1837), no. 12, pp. 137-148. Partial English translation (from the French) in E. Straker, *Wealden Iron* (London, 1931).

9. Caesalpinus, A. *De Metallicia Libri Tres.* Rome, 1596.

10. Libavius, A. *Alchymia . . . Recognita, Emendata, et Aucta. Commentariorum Alchymiae Andrae Libavii Med.* Frankfurt, 1606.

11. Mathesius, J. *Sarepta oder Bergpostill.* Nürnberg, 1562.

12. Monardes, N. *Dialogo del Hierro y de Sus Grandezas.* Seville, 1571. German translation by J. Gesner, *Ein nützlich und lustig Gespräch von Stahl und Eisen* (Leipzig, 1615). English translation by J. Frampton, "The Dialogue of Iron," in Monardes, *Joyfull News out of the New Founde Worlde* (London, 1577; reprinted 1925).

13. Perez de Vargas, B. *De Re Metallica.* Madrid, 1569.

14. Porta, G. B. della. *Magia Naturalis.* Naples, 1589. English translation (London, 1658; reprinted 1957).

15. Schwenckfeldt, G. *Stirpium et Fossilium Silesiae Catalogus.* Leipzig, 1600.

16. *Von Stahel und Eysen. . . .* Nürnberg, 1532. English translation in C. S. Smith, ed., *Sources for the History of the Science of Steel* (Cambridge, Mass., 1968).

17. Sturtevant, S. *Metallica.* London, 1612 (reprinted 1855, 1858).

18. Taubmanus, F. *Ardea Kochticziorum. Ad Generosum . . . Andream Kochtitzki, Musagetam Eminentiss[imum].* Wittenberg, 1612. Second edition in *Melodaesiae sive epulum Musaeum* (Leipzig, 1615).

19. [Winter, K.] *Fabrilium Silesiae Officinarum Fodinarumque Descriptio et Denotatio Brevis.* Frankfurt, 1556. Third edition in N. Henelius, *Silesiographia Renovata* (Vratislava, 1704).

c. Secondary Works on the History of Iron and Steel and Other Auxiliary Sources

1. Aitchison, L. *A History of Metals.* 2 vols. London, 1960.

2. Baranowski, I. *Przemysł polski w XVI wieku* [Polish industry in the 16th century]. Warsaw, 1919.

3. Barycz, H., and Zathey, J., eds. *Album Studiosorum Universitatis Cracoviensis.* Krakow, 1950: vol. 4, p. 104.

4. Bastyr, W., and Paszkowski, E. *Anglo-Polish Translation of Workshop Terms.* London, 1946: p. 34.

5. Beck, L. *Geschichte des Eisens in seiner technische und kulturgeschichtlichen Beziehung.* 5 vols. Braunschweig, 1891-1903.

6. Brückner, A. *Tysiąc lat kultury polskiej* [A thousand years of Polish culture]. 2 vols. Paryż, 1955.

7. Brückner, A. *Dzieje kultury polskiej* [The history of Polish culture]. 3 vols. Warsaw, 1958.

8. Buchwald-Pelcowa, P. Inwentarz biblioteki Macieja Grabskiego [Inventory of the library of Maciej Grabski]. *Rocznik Biblioteki Narodowej,* vol. 2 (1966), pp. 318-337.

9. Fedorenko, P. K. Rudni levoberežnoi Ukrainy v XVI-XVIII v. [Bloomeries of the Left-Bank Ukraine in the 16th through 18th centuries]. Moscow, 1960.

10. Forbes, R. J. *Metallurgy in Antiquity.* Leiden, 1950.

11. Forbes, R. J. *Studies in Ancient Technology.* Leiden, 1964: vol. 9.

12. Formanowicz, L. Biblioteka kapitulna w Gnieźnie [Library of the Gniezno Chapter House]. In A. Wierczyński, ed., *Biblioteki wielkopolskie i pomorskie* [Libraries in Greater Poland and Pomerania]. Poznań, 1929.

13. François, J. *Essai sur l'élaboration du minerai de fer dans le traitement à la catalane.* Foix, 1838.

14. Fullon, A. *O vydelkie želaza v syrodutnych pecach* [On the production of iron in bloomery furnaces]. St. Petersburg, 1819.

15. Grabig, H. *Die mittelalterliche Eisenhüttenindustrie der Niederschlesisch-Lausitzer Heide und ihre Wasserhämmer.* Breslau, 1937.

16. Grier, W. *The Mechanic's Calculator.* 2nd ed. Glasgow, 1835.

17. Haiek, V. z Liboczan, *Kronyka czeska* [Czech chronicle]. Prague, 1541.

18. Heller, F. F. *Lob des Eisens.* Düseldorf, 1965.

19. Johannsen, O. *Geschichte des Eisens.* 3rd ed. Düsseldorf, 1953.

20. Kolčin, B. A. Čornaia metallurgia i metalloobrabotka v drevnei Rusi [Iron metallurgy and working in Old Ruthenia]. *Materialy i Issledovania po Archeologii SSSR,* vol. 32 (1954), p. 5-257.

21. Kořan, J. *Staré české železářství* [Old Czech iron metallurgy]. Prague, 1946.

22. Kořan, J. K periodisaci dějin českého hornictví [On the division of the history of Czech mining into periods]. *Sbornik pro Hospodarské a Socialni Dějiny,* vol. 1 (1946).

23. Krzyżanowski, J. *Historia literatury polskiej od średniowiecza do XIX w.* [The history of Polish literature from the Middle Ages to the 19th century]. Warsaw, 1953.

24. Krzyżanowski, J. *Historia literatury poslkiej. Alegoryzm-preromantyzm* [The history of Polish literature. Allegorism to preromanticism]. Warsaw, 1964.

25. Kuraś, S. Materiały do górnictwa i hutnictwa z Archiwów Metropolitalnego i Kapitulnego w Krakowie 1479-1640 [Material on mining and metallurgy from the Metropolitan and Chapter Archives in Krakow, 1479-1640]. *Studia z Dziejów Górnictwa i Hutnictwa,* vol. 3 (1959), pp. 307-308.

26. Miczulski, S. Początki i rozwój pierwszego ośrodka wielkopiecowego na ziemiach Polski [The beginnings and development of the first blast-furnace center in Polish territory]. *Wiadomości Hutnicze,* vol. 18 (1962), no. 7-8, pp. 227-229.

27. Miczulski, S. Introduction of the blast-furnace process in the Polish metallurgical industry. In *Études d'histoire de la science et de la technique* (Wrocław-Warsaw-Krakow, 1968), pp. 255-261.

28. Musioł, L. *Materiały do dziejów Wielkich Katowic, 1299-1799* [Materials on the history of Greater Katowice, 1299-1799]. Katowice, 1936.

29. Musioł, L., and Płuszczewski, S. Wykaz zakładów dawnego hutnictwa żelaza na Górnym Śląsku od XIV do połowy XIX w. [A list of former ironworks in Upper Silesia from the 14th through the mid-19th centuries]. *Studia z Dziejów Górnictwa i Hutnictwa,* vol. 5 (1961), pp. 7-87.

30. Osiński, J. *Opisanie polskich żelaza fabryk* [Descriptions of Polish ironworks]. Warsaw, 1782.

31. Pazdur, J., ed. *Zarys dziejów górnictwa na ziemiach polskich* [Outline of the history of mining in Polish territory]. 2 vols. Katowice, 1960-1961.

32. Pelc, J. *Ceny w Krakowie w latach 1369-1600* [Prices in Krakow in the period 1369-1600]. Lwow, 1935.

33. Percy, J. *Metallurgy,* vol. 4: *Iron and Steel.* London, 1884.

34. Piaskowski, J. Metalurgia w XVI wieku w świetle dzieła Agricoli *De re metallica* [Metallurgy in the 16th century in the light of Agricola's work *De re metallica*]. In *Georgius Agricola, 1494-1555* (Wrocław, 1957), pp. 179-185.

35. Piaskowski, J. O pierwotnym sposobie otrzymywania stali u średniowiecznych Słowian [On the original method of steel production among medieval Slavs]. *Hutnik,* vol. 31 (1964), no. 12, pp. 378-383.

36. Piaskowski, J. Dul w dawnej polskiej literaturze technicznej i jego interpretacja [The *dul* in the old Polish technological literature and its interpretation]. *Kwartalnik Historii Kultury Materialnej,* vol. 13 (1965), no. 1, pp. 55-63.

37. Piaskowski, J. Zagadnienie ciągłości rozwoju hutnictwa żelaza na ziemiach polskich w starożytności i we wczesnym

średniowieczu [The problem of the continuity of iron metallurgy development in the Polish territories in antiquity and in early medieval times]. *Roczniki Dziejów Społecznych i Gospodarczych,* vol. 32 (1971), pp. 1-29.

38. Piernikarczyk, J. *Historia górnictwa i hutnictwa na Górnym Śląsku* [The history of mining and metallurgy in Upper Silesia]. 2 vols. Katowice, 1933.

39. Piwarski, K. *Historia Śląska* [The history of Silesia]. Katowice-Wrocław, 1947.

40. Pleiner, R. *Základy slovanského železářskeho hutnictví v českych zemich* [Of Slav iron metallurgy works in Czech territory]. Prague, 1958.

41. Popiołek, F. *Dzieje hutnictwa żelaznego na ziemiach polskich* [The history of iron metallurgy in Polish territory]. Katowice-Wrocław, 1947.

42. Radwan, M. Kiedy powstał i jak wyglądał pierwszy wielki piec w Polsce [How the first blast furnace arose in Poland and what it was like]. *Hutnik,* vol. 8 (1936), no. 2, pp. 71-75.

43. Radwan, M. *Rudy, kuźnice i huty żelaza w Polsce* [Smelting houses, bloomeries, and ironworks in Poland]. Warsaw, 1963.

44. Rospond, S. *Zabytki języka polskiego na Śląsku* [Remaining traces of the old Polish language in Silesia]. Wrocław-Katowice, 1948: pp. 109-113.

45. Rospond, S. *Nazwiska Ślązaków* [Names of Silesians]. Opole, 1960: pp. 77-78.

46. Rutkowski, J. *Historia gospodarcza Polski* [An economic history of Poland]. 3rd ed. 2 vols. Poznań, 1947.

47. Rybarski, R. *Handel i polityka handlowa Polski w XVI stuleciu* [Trade and the trade policies of Poland in the 16th century]. 2 vols. Poznań, 1928-1929.

48. Schubert, H. R. *History of the British Iron and Steel Industry.* London, 1957.

49. Singer, G., et al., eds. *A History of Technology.* 5 vols. Oxford, 1958.

50. Strumilin, S. G. *Istoria černoi metallurgii w SSSR* [History of iron metallurgy in the Soviet Union]. Moscow, 1954.

51. Swank, J. M. *History of the Manufacture of Iron and Steel in All Ages.* Philadelphia, 1884.

52. Sygański, J. Z dawnych metryk Kościoła Mariackiego w Krakowie [From old registers of St. Mary's Church in Krakow]. *Miesięcznik Heraldyczyn,* no. 3 (1910), p. 116.

53. Tomaszewski, E. Ceny w Krakowie w latach 1601-1795 [Prices in Krakow in the period 1601-1795]. Lwów, 1934.

54. Trepka, W. N. *Liber Generationis Plebeanorum.* Wrocław, 1963.

55. Wertime, Th. A. *The Coming of the Age of Steel.* Chicago, 1961.

56. Wojtas, M., ed. *Akta wizytacji dekanatu pszczyńskiego i bytomskiego z r. 1598, dokonanej w roku 1598 z polecenia Jerzego Kardynła Radziwiłła* [Documents of inspection of

Pszczyna and Bytom deaneries from 1598, taking place in 1598 by order of Jerzy Cardinal Radziwiłł]. Katowice, 1939.

57. Zaleski, W. *Dzieje górnictwa i hutnictwa na Górnym Śląsku do roku 1806* [The history of mining and metallurgy in Upper Silesia up to the year 1806]. Madryt, 1967.

58. Zientara, B. Ze studiów nad historią techniki polskiego hutnictwa żelaznego do XVII wieku [From studies of the history of the technology of Polish iron metallurgy up to the 17th century]. *Przegląd Historyczny,* vol. 43 (1952), no. 2, pp. 195-242.

59. Zientara, B. *Dzieje małopolskiego hutnictwa żelaznego XIV-XVII w.* [The history of iron metallurgy in Little Poland in the 14th to 16th centuries]. Warsaw, 1954.

60. Zivier, E., ed. *Acten und Urkunden zur Geschichte des schlesischen Bergwessens. Österreichische Zeit.* Katowice, 1900.

61. Zivier, E. Zur ältesten Geschichte der Eisenindustrie in Oberschlesien. *Oberschlesien,* vol. 1 (1902).

62. Zivier, E. *Entwicklung und Bedeutung der oberschlesischen Eisenindustrie.* Katowice, 1913.

d. A Selection of Studies on the Officina Ferraria

1. Bizoń, F. Błędy drukarskie pierwodruku polskiego [Misprints in the first Polish edition (of the *Officina Ferraria*)]. In the 1962 edition of the *Officina Ferraria* (a1962), pp. LXXXVIII-XCIII.

2. Bizoń, F. Indeks wyrazów i form [Index of words and forms (in the *Officina Ferraria*)]. In the 1962 edition (a1962), pp. 222-316.

3. Haertel, E. Recenza wydania "Officina ferraria" z 1933 [Review of 1933 edition of *Officina Ferraria*]. *Jahrbücher für Geschichte und Kultur der Slaven,* vol. 11 (1935), pp. 158-160.

4. Haertel, E. Bemerkungen und Ermittlungen zu des Walenty Roździeński "Officina ferraria." *Zeitschrift für slavische Philologie,* vol. 12 (1935), pp. 110-119.

5. Haertel, E. Bergbau und Hütteindustrie Oberschlesiens in alter Zeit. Zu den bisher unbekannten Bericht von Roździeński aus dem Jahre 1612. *Der Oberschlesier,* vol. 17 (1935), no. 4, pp. 198-203.

6. Haertel, E. Die Bodenschätze im Riesengebirge nach einer alten polnischen Dichtung. *Wanderer im Risengebirge,* vol. 55 (1935), no. 7, pp. 113-115.

7. Haertel, E. Andreas Freiherr von Kochtizky, ein Mäzen Oberschlesiens in die Zeit des Dreissigjährigen Krieges. *Der Oberschlesier,* vol. 18 (1936), no. 9, pp. 487-493, and no. 11, pp. 578-587.

8. Haertel, E. Ein Nachkomme der meissener Famillie Herzig schreibt in 1612 über die oberschlesische Hüttenindustrie. *Der Oberschlesier,* vol. 19 (1957), no. 5, pp. 156-160.

9. Haertel, E. Recenzja wydania "Officina ferraria" z 1936 [Review of 1936 Edition of *Officina Ferraria*]. *Zeitschrift für slavische Philologie*, vol. 14 (1937), pp. 471-473.

10. Haertel, E. Officina ferraria. Eine praktische Hüttenkunde aus dem Jahre 1612. *Stahl und Eisen*, vol. 60 (1940), no. 11, pp. 233-235.

11. Haertel, E. Aus der Frühzeit des oberschlesischen Hüttenwesens. *Beiträge zur Geschichte der Technik und Industrie*, vol. 29 (1940), pp. 163-164.

12. Haertel, E. Walenty Roździeński's polnische Bergmannsgedichte von Jahre 1612 und ihre Beziehungen zur deutschen Kultur. *Veröffentlichungen des slavischen Instituts and der Friedrich-Wilhelms-Universität Berlin*, vol. 27 (1940), pp. 1-65.

13. Jarosz, A. Ze zagadnień świadomości narodowej i słowiańskiej W. Roździeńskiego [On problems of national and Slav consciousness of W. Roździeński]. *Zeszyty Naukowe Wyższej Szkoły Pedagogicznej w Katowicach*, 1965, no. 24, pp. 95-111.

14. Jarosz, A. Osiągnięcia i perspektywy badań nad "Officina ferraria" [Achievements and perspectives of the investigations upon the *Officina Ferraria*]. *Zaranie Śląskie*, vol. 29 (1966), no. 2, pp. 213-233.

15. Jarosz, A. Nowe studia o "Officina ferraria" [New studies on the *Officina Ferraria*]. *Zaranie Śląskie*, vol. 30 (1967), no. 1, pp. 61-71.

16. Jarosz, A. Uwagi o epickości "Officina ferraria" W. Roździeńskiego [Notes on the Epicism of the *Officina Ferraria* of W. Roździeński]. *Zeszyty Naukowe Wyższej Szkoły Pedagogicznej w Katowicach*, 1967, no. 4, pp. 41-59.

17. Jarosz, A., and Lubaś, W. Historia i znaczenie nazwy osobowej Roździeński [The history and significance of the personal name Roździeński]. *Zeszyty Naukowe Wyższej Szkoły Pedagogicznej w Katowicach*, 1965, no. 24, pp. 115-125.

18. Jarosz, A., and Ocieczek, R. Roździeńściana. *Pamiętnik Literacki*, vol. 55 (1964), nos. 3-4, pp. 233-255.

19. Kaminsky, F. Kampf um Eisen in Oberschlesien an der Schwelle des 17 Jahrhunderts. *Deutsche Monatshefte, Zeitschrift für Geschichte und Gegenwart des Ostdeutschtums*, vol. 7 (1940/1941), no. 17, pp. 390-398.

20. Klich, E. Krytyczne uwagi o drugim wydaniu "Officina ferraria" [A critical commentary on the second edition of the *Officina Ferraria*]. *Prace Filologiczne*, vol. 17 (1937), pp. 157-171.

21. Kuczewski, W. Poemat z r. 1612 o rudach, hutach i kuźnikach na Śląsku i w Polsce [A poem from the year 1612 on ores, ironworks, and forgemasters in Silesia and Poland]. *Hutnik*, vol. 5 (1933), pp. 255-261.

22. Kudera, J. *Walenty Roździeński, sławny mistrz i autor książki o hutnictwie śląskim w XVII w.* [Walenty Roździeński, famous master and author of a book on Silesian metallurgy in the 17th century]. Mysłowice, 1938.

23. Kuglin, J. Szymon Kempini i jego druk "Officina

ferraria" [Szymon Kempini and his publication of the *Officina ferraria*]. In S. Rospond, ed., *Rożdzieńsciana* (Wrocław-Warsaw-Krakow, 1965), pp. 247-258.

24. Łanowski, J. Czapla Kochcickich Rożdzieńścianum [The heron in the coat of arms of the Kochcicki family according to Rożdzieński]. *Pamiętnik Literacki,* vol. 56 (1965), no. 3, pp. 149-166.

25. Łanowski, J., and Myśliwiec, H. Walentego Rożdzieńskiego "erudycja klasyczna" [Walenty Rożdzieński's classical erudition]. In S. Rospond, ed., *Rożdzieńsciana* (Wrocław-Warsaw-Krakow, 1965), pp. 142-184.

26. Ligęza, J. Tradycja jako źródło "Officina ferraria" [Tradition as a source in the *Officina Ferraria*]. In S. Rospond, ed., *Rożdzieńsciana* (Wrocław-Warsaw-Krakow, 1965), pp. 247-258.

27. Łysik, S. Jurga Kleparski, śląski kuźnik średniowieczny. Próba wyjaśnienia pochodzenia i nazwiska [Jurga Kleparski, the medieval Silesian forgemaster. An attempt to explain his descent and name]. *Zaranie Śląskie,* vol. 27 (1964), no. 1, pp. 88-89.

28. Mágr, A. S. O českém železářstvi v polsém poematu z r. 1612 [On Czech iron smelting in a Polish poem of 1612], *Slezaký Sborník,* 1952, no. 9, pp. 129-135.

29. Miczulski, S. Nowe wiadomości o Walentym Rożdzieńskim [New information on Walenty Rożdzieński]. In *Komunikat przedstawiony na Sesji Zespołu Historii Polskiej Techniki Hutniczej i Odlewniczej Polskiej Akadmii Nauk* (Kielce, 1973).

30. Musioł, L. Z źródłowych dociekań nad nazwami oraz nazwiskami w poemacie Rożdzieńskiego "Officina ferraria" [Research on the sources of personal and place names in Rożdzieński's poem *Officina Ferraria*]. *Zaranie Śląskie,* vol. 11 (1935), no. 4, pp. 292-294.

31. Myśliwiec, H. Nieznany Isatius jako źródło W. Rożdzieńskiego [The unknown Isatius as a source in W. Rożdzieński]. *Eos,* vol. 54 (1964), pp. 360-366.

32. Myśliwiec, H. Rożdzieńścianum: f. L_2 vv. 13-16. *Pamiętnik Literacki,* vol. 55 (1964), no. 1-2, pp. 451-453.

33. Ocieczek, R. Główne kierunki recepcji "Officina ferraria" W. Rożdzieńskiego [Main trends of reception of the *Officina Ferraria* of W. Rożdzieński]. *Zeszyty Naukowe Wyższej Szkoły Pedagogicznek w Katowicach,* 1965, no. 24, pp. 49-93.

34. Ocieczek, R. Kształt literacki listu dedykacyjnego "Officina ferraria" Walentego Rożdzieńskiego [The literary form of the dedicatory letter in the *Officina Ferraria* of Walenty Rożdzieński]. *Zeszyty Naukowe Wyższej Szkoły Pedagogicznej w Katowicach,* 1967, no. 40, pp. 61-73.

35. Ocieczek, R. "Officina ferraria" w siedemnastowicznym księgozbiorze Jana Kazimierza Grabskiego [*Officina Ferraria* in the 17th century library belonging to Jan Kazimierz Grabski]. *Zaranie Śląskie,* vol. 31 (1968), no. 2, pp. 263-273.

36. Pazdur, J. Przyczynek do interpretacji "Officyny" W. Rożdzieńskiego [A contribution to the interpretation of the

Officina of W. Roździeński]. *Kwartalnik Historii Kultury Materialnej,* vol. 11 (1963), no. 3-4, pp. 591-599.

37. Piasecki, W., Jarecka, H., Jarosz, A., Ocieczek, R., and Wawrzykiewicz, J. Bibliografia Roździeńskiego i jego dzieła [Bibliography of Roździeński and his work]. In S. Rospond, ed., *Roździeńsciana* (Wrocław-Warsaw-Krakow, 1965), pp. 6-63.

38. Piaskowski, J. "Officina ferraria" Walentego Roździeńskiego—polski poemat hutniczy z 1612 roku [*Officina Ferraria* of Walenty Roździeński—Polish metallurgical poem of 1612]. *Wiadomości Hutnicze,* vol. 25 (1969), no. 4, pp. 105-109.

39. Piaskowski, J. Ze studiów nad poematem hutniczym W. Roździeńskiego "Officina ferraria" [From the studies on W. Roździeński's metallurgical poem *Officina Ferraria*]. *Kwartalnik Historii Nauki i Techniki,* vol. 15 (1970), no. 1, pp. 31-57.

40. Płuszczewski, S. Z dziejów polskiego słownictwa hutniczego [Concerning the history of Polish metallurgical terminology]. *Wiadomości Polskiego Komitetu Normalizacyjnego,* vol. 20 (1952), no. 7, pp. 526-529.

41. Pollak, R. Staropolski poemat o górnikach i hutnikach [An old Polish poem on miners and smelters]. *Przegląd Współczesny,* vol. 12 (1933), no. 138, pp. 112-124.

42. Pollak, R. Z dziejów hutnictwa w Polsce [Concerning the history of metallurgy in Poland]. *Hutnik,* vol. 8 (1936), no. 11, pp. 457-460.

43. Pollak, R. Róździeńściana. Echa poematu Roździeńskiego o "szlachetnem dziele żeleznem" [Echoes of Roździeński's poem on "The noble craft of ironwork"]. *Zaranie Śląskie,* vol. 12 (1936), no. 2, pp. 101-104.

44. Pollak, R. Krytyczne uwagi nad krytycznymi uwagami o wydaniu "Officina ferraria" [Critical comments on the critical comments on the publication of the *Officina Ferraria*]. *Zaranie Śląskie,* vol. 14 (1938), no. 1, pp. 44-47.

45. Pollak, R. Roździeńściana. W kręgu poematu Walentego Roździeńskiego [In the sphere of Walenty Roździeński's poem]. *Zaranie Śląskie,* vol. 14 (1938), pp. 240-245.

46. Pollak, R. Roździeńścianum. III. *Zaranie Śląskie,* vol. 17 (1946), no. 3/4, pp. 119-121.

47. Pollak, R. Roździeński i "Sarepta" Methesiusa [Roździeński and the *Sarepta* of Mathesius], *Zaranie Śląskie,* vol. 18 (1947), no. 4, pp. 196-198.

48. Pollak, R. Roździeńściana. IV. *Zaranie Śląskie,* vol. 18 (1947), no. 4, pp. 199-204.

49. Pollak, R. Powojenne Roździeńściana [Postwar Roździeńściana]. *Pamiętnik Literacki,* vol. 45 (1954), no. 1, pp. 223-229.

50. Pollak, R. Wstęp [Introduction]. In the 1962 edition of the *Officina Ferraria* (a1962), pp. VII-XVIII.

51. Pollak, R. "Huta" Roździeńskiego a pokrewne utwory obce [Roździeński's "smelting house" and kindred foreign works]. In S. Rospond, ed., *Roździeńsciana* (Wrocław-Warsaw-Krakow, 1965), pp. 64-75.

52. Pollak, R. Na marginesach ostatnich "Roździeńscianow" [On the margins of the last "Roździeńsciana"]. *Pamiętnik Literacki*, vol. 56 (1965), no. 3-4, pp. 231-233.

53. Pollak, R., and Rospond, S. Objaśnienia [Explanations]. In the 1962 edition of the *Officina Ferraria* (a1962), pp. 187-226.

54. Radwan, M. Świadek wiarygodny [A reliable witness]. In the 1962 edition of the *Officina Ferraria* (a1962), pp. LXIII-LXXXVII.

55. Radwan, M. Wiedza górnicza i hutnicza Walentego Roździeńskiego, w 350-lecie "Officina ferraria" [Walenty Roździeński's knowledge of mining and metallurgy, on the 350th anniversary of the *Officina Ferraria*]. *Kwartalnik Historii Nauki i Techniki*, vol. 8 (1963), no. 4, pp. 491-501.

56. Radwan, M. O kuźnicach bez kunsztów wodnych i o kuźnikach polskich na Białorusi [On forges without water-driven machinery and Polish forgemasters in White Ruthenia]. In S. Rospond, ed., *Roździeńsciana* (Wrocław-Warsaw-Krakow, 1965), pp. 234-246.

57. Rospond, S. W. Roździeński a J. Kochanowski. *Język Polski*, vol. 41 (1961), no. 4, pp. 295-302.

58. Rospond, S. Uwagi o języku utworu [Notes on the language of the work]. In the 1962 edition of the *Officina Ferraria* (a1962), pp. XVIII-LXII.

59. Rospond, S. Walenty Roździeński a Jan Kochanowski. Rymotwórstwo w "Officina ferraria" [Walenty Roździeński and Jan Kochanowski. Versification in the *Officina Ferraria*]. In S. Rospond, ed., *Roździeńsciana* (Wrocław-Warsaw-Krakow, 1965), pp. 73-141.

60. Rospond, S. Polski "skarbnik" a niemiecki Rübezahl [The Polish specter guardian of treasure and the German "Rübezahl"]. In S. Rospond, ed., *Roździeńsciana* (Wrocław-Warsaw-Krakow, 1965), pp. 254-258.

61. Rospond, S. Autentyzm toponimii śląskiej w "Officina ferraria" [The authenticity of Silesian toponymy in the *Officina Ferraria*]. *Zaranie Śląskie*, vol. 28 (1965), pp. 326-335.

62. Rospond, S. Jeszcze o Jurdze Kleparskim [Some more words about Jurga Kleparski]. *Zaranie Śląskie*, vol. 28 (1965), no. 3, pp. 708-710.

62a. Rospond, S., ed. *Roździeńsciana, Studia o Walentym Roździeńskim autorze "Officina ferraria" z 1612 r.* Wrocław-Warsaw-Krakow, 1965.

63. Steffen, W. Antike Quellen der "Officina Ferraria" von Walenty Roździeński. *Eos*, vol. 52 (1962), no. 2, pp. 351-359.

64. Waszkis, H. Trzynastogłowiec w "Officina ferraria" [The thirteen-foot verses in the *Officina Ferraria*]. In S. Rospond, ed., *Roździeńsciana* (Wrocław-Warsaw-Krakow, 1965), pp. 259-266.

65. Vasmer, M. Die Reime in W. Roździeński's "Officina ferraria." *Zeitschrift für slavische Philologie*, vol. 12 (1935), pp. 357-358.

66. Zaremba, J. Twórca "Officina ferraria" a grupa zawodowa górników górnośląskich [The author of *Officina Ferraria*

and the professional group of Upper Silesian miners]. In S. Rospond, ed., *Roździeńsciana* (Wrocław-Warsaw-Krakow, 1965), pp. 183-205.

67. Zaremba, J. Poemat Roździeńskiego o hutnikach śląskich [Poem of W. Roździeński on Silesian metallurgists]. Polska Akademia Nauk, Oddział w Krakowie, Seria "Nauka dla Wszystkich," no. 63, Krakow, 1968.

POLISH METALLURGICAL TERMS

English	Polish
anvil	kowadło, kowalno, nakowalno
arms	ramiona
blast furnace	wielki piec
bellows	miechy
bellow cams	nogi, dyszynogi
bloom	łupa
bloomery forge	kuźnica dymarska
bloomery hearth	piec dymarski
bloomery iron	żelazo łupne
bushes	buksze
charcoal	węgiel drzewny
charcoal kiln	mielerz
clay slurry	hamraj
crowbar	stáchel
cutter, shears	szrotyzn, przecinak
finery, Italian forge	fryszerka, kuźnica włoska
hammer scale	hamerszlaka, młotowina, zendra
hook, swage	otuł, hak
hot set	odsadzka, hesprys
instruments for power-driven forge	strój, nastroj, kunszty
monk	mnich
piece of bloom	szerbel
piercer	probój, zynkiesz
recoil beam	ryttel, odbijak
slag notch	lachor
sleeve of the hammer helm	helza, elza
sluice	upust
smith's hearth	piecek, piec kowalski
stamp mill	stępa, tłuczka
steel	ocel, oczyl, stal
Styrian furnace	piec styryjski
tongs	klyszcze
tuyere	forma
tuyere-hook	formyzen, formizel

English	Polish
water hammer	mechaniczny młot wodny
waterwheel	koło wodne
waterwheel shaft	wał koła wodnego

See also pp. 24, 25, 27.

INDEX